The Future of Liquified Natural Gas in a Decarbonising World

This book analyses the role of liquified natural gas in a decarbonising world and presents the most significant energy-transition options and implications for the liquified natural gas industry.

Major investments and developments in technology have been made in recent years in an attempt to meet global demand, but energy systems require radical new pathways to meet climate goals in line with the Paris Agreement. This book explores the role of liquified natural gas in the context of the global energy transition, arguing that liquified natural gas has a role to play in terms of resources, the gas market, energy-transition dynamics/regime status, and geopolitical powers. Using a bespoke meta-framework grounded in institutional theory and case studies, the book examines how institutional, political, and resource characteristics affect liquified natural gas use. The book also explores implications for liquified natural gas exporters in the context of the energy transition and analyses the characteristics of liquified natural gas compared with pipeline gas. The multiple case study approach examines the role of natural gas in Japan, the UK, and China, three countries in different stages of the energy transition, to determine potential pathways for exporters. Utilising a multi-method procedure for data collection, including data analysis, in-depth interviews, and direct observations, the book concludes with findings on the potential role of liquified natural gas in various future stages of the energy transition.

Written by an industry expert, this book will be of value to students, researchers, and academics interested in energy studies, decarbonisation studies, and environmental studies more broadly.

Omran Al-Kuwari is an international executive with over 20 years of experience in various industries, including clean energy, hydrocarbons, education, and R&D, in both private and public sectors. After joining LNG supplier Qatargas in 1999, Al-Kuwari led several large-scale projects for its different global joint ventures and other Qatar Petroleum affiliates in Doha, Japan, and the UK.

Routledge Explorations in Energy Studies

Local Energy Communities
Emergence, Places, Organizations, Decision Tools
Edited by Gilles Debizet, Marta Pappalardo & Frédéric Wurtz

Building Resilient Energy Systems
Lessons from Japan
Jennifer F. Sklarew

Regulatory Support for Off-Grid Renewable Electricity
Edited by Ngozi Chinwa Ole, Eduardo G. Pereira, Peter Kayode Oniemola and Gustavo Kaercher Loureiro

Northern Indigenous Community-Led Disaster Management and Sustainable Energy
Ranjan Datta, Margot Hurlbert and William Marion

Energy Policy Design in the South-Eastern Mediterranean Basin
A Roadmap to Energy Efficiency
Bertug Ozarisoy and Hasim Altan

Living with Energy Poverty
Perspectives from the Global North and South
Edited by Paola Velasco-Herrejón, Breffní Lennon and Niall Dunphy

Transitioning Fossil-Based Economies
Sustainable Strategies for Energy Change
Hassan Qudrat-Ullah

The Future of Liquified Natural Gas in a Decarbonising World
Omran Al-Kuwari

For more information about this series, please visit: www.routledge.com/Routledge-Explorations-in-Energy-Studies/book-series/REENS

The Future of Liquified Natural Gas in a Decarbonising World

Omran Al-Kuwari

Routledge
Taylor & Francis Group
LONDON AND NEW YORK

earthscan
from Routledge

First published 2025
by Routledge
4 Park Square, Milton Park, Abingdon, Oxon OX14 4RN

and by Routledge
605 Third Avenue, New York, NY 10158

Routledge is an imprint of the Taylor & Francis Group, an informa business

British Library Cataloguing-in-Publication Data
A catalogue record for this book is available from the British Library

ISBN: 978-1-032-89890-2 (hbk)
ISBN: 978-1-032-89891-9 (pbk)
ISBN: 978-1-003-54514-9 (ebk)

DOI: 10.4324/9781003545149

Typeset in Times New Roman
by Apex CoVantage, LLC

Contents

Figures

Tables

Abbreviations

BCM	billion cubic metres
BEIS	Department for Business, Energy and Industrial Strategy (UK)
BP	British Petroleum
BRI	Belt and Road Initiative (China)
CCC	Committee on Climate Change
CCS	carbon capture and storage
CCUS	carbon capture, utilisation, and storage
CNEA	China Nuclear Energy Agency
CNOOC	China National Offshore Oil Corporation
CNPC	China National Petroleum Company
CPC	Communist Party of China
DPJ	Democratic Party of Japan
EIA	US Energy Information Administration
ETS	emissions trading scheme
EU	European Union
FIT	feed-in tariff
FLNG	floating liquified natural gas
FSRU	floating storage regasification unit
FYP	Five-Year Plan (China)
GDP	gross domestic product
GECF	Gas Exporting Countries Forum
GHG	greenhouse gas
Gt	gigaton (mass)
GWP	global warming potential
IEA	International Energy Agency
IEEJ	Institute of Energy Economics Japan
IGU	International Gas Union
IMO	International Maritime Organization
INDC	intended nationally determined contribution
IPCC	Intergovernmental Panel on Climate Change
IRENA	International Renewable Energy Agency
LDP	Liberal Democratic Party (Japan)
LNG	liquified natural gas

METI	Ministry of Economy, Trade and Industry (Japan)
MMBTu	1 million British thermal units
MoE	Ministry of Energy (Japan)
Mt	megaton (mass)
MTPA	million tonnes per annum
NDC	nationally determined contribution
NDRC	National Development and Reform Commission (China)
NGO	non-governmental organisation
NOC	national oil company
OPEC	Organization of the Petroleum Exporting Countries
$PM_{2.5}$	particulate matter
R&D	research and development
SEP	strategic energy plan (Japan)
SNG	synthetic natural gas
SOE	state-owned entity
TCF	trillion cubic feet
TWh	terawatt hours
UN	United Nations
UNFCCC	The United Nations Framework Convention on Climate Change

Acknowledgements

I wish to express my profound gratitude to the mentors who have guided me throughout my career and journey. Some include Athar Masood, Ghanim Al-Kuwari, Ali Al-Hammadi, Faisal Al-Suwaidi, Dr Tidu Maini, and particularly Professor Michael Grubb and Professor Albert Bressand. Their exceptional wisdom, generosity, and unwavering support have been invaluable, and I am deeply fortunate to have had the opportunity to learn from them. Their influence has shaped not only my professional and academic pursuits but also my broader understanding of the complexities within the field.

I am also immensely thankful to the esteemed institutions that have supported and nurtured my development over the years, including Qatar Energy, ExxonMobil, Qatar Foundation, and the Institute for Sustainable Resources at UCL. Their commitment to fostering my growth has been instrumental in shaping the direction and substance of my work.

Over the past 25 years, I have been privileged to observe the evolution of the energy transition from multiple vantage points – academia, the private and public sectors, investment, and policy development. This multifaceted experience has not only deepened my understanding of the global energy landscape but has also inspired me to contribute to the ongoing dialogue on sustainability. I hope that this book will offer valuable insights into the critical role of LNG in the energy transition and contribute meaningfully to the broader debate on this pivotal issue.

Finally, I extend my heartfelt thanks to my family and dear friends. Your unwavering support and encouragement have been my foundation throughout this journey, and I am profoundly grateful for your presence in my life. Thank you!

1 Introduction

A global energy transition is currently taking place, driven primarily by the need to combat climate change. The Intergovernmental Panel on Climate Change (IPCC) has concluded that the current trajectory of global greenhouse gas (GHG) emissions is not consistent with limiting global warming to below 1.5°C or 2°C relative to pre-industrial levels, a threshold that could lead to severe economic damage and instability in the coming decades. Fossil fuel combustion, industry, transport, and electricity production contribute to approximately 80% of global GHG emissions (Al-Kuwari, 2023; Al-Yafei et al., 2021). Energy systems must therefore decarbonise at dramatic rates to move towards a more sustainable environmental development path and cater for population and economic growth in many parts of the world. Policymakers face the difficult task of moving towards more environmentally sustainable pathways, while balancing the need to develop and grow their economies.

Natural gas, a fuel with more environmental credentials than other fossil fuels, is a "transition fuel" that can support the low-carbon transition by promoting fuel-switching and supporting hard-to-abate sectors until large-scale electrification with renewable resources and other solutions such as large-scale batteries and hydrogen are developed and deployed (Al-Yafei et al., 2021; Kim et al., 2023; Lin & Brooks, 2021). Natural gas has been the fastest-growing fossil fuel over the past few decades, propelled by the environmental benefits it has over other traditional fossil fuel sources and by an abundance of new supply that has arisen as a result of technological advances in gas extraction. Nonetheless, the role of natural gas as a bridge to a low-carbon world remains unclear. It is forecast to increase in coming decades, with a longer-trend shift towards lower-carbon sources depending on the availability of carbon capture and other decarbonisation technologies (Al-Kuwari, 2023; Lin & Brooks, 2021).

Liquified natural gas (LNG) is the fastest-growing segment within natural gas (Energy Institute, 2024). However, relatively little attention has been paid to its role in the energy transition. LNG may well grow and exceed the pipeline gas trade by 2025 according to the International Energy Agency (IEA) because of its commercial and technical flexibility and the significant increases in unconventional gas supply in the United States (Selden, 2020). Although literature on the role of natural gas in the energy transition is extensive, it focuses primarily on debating the

DOI: 10.4324/9781003545149-1

merits of natural gas as a transition fuel (Al-Yafei et al., 2021). Accordingly, LNG has often been treated largely as a form of natural gas transportation rather than as a stand-alone category, technology, or system worthy of its own study. There is a difference between LNG and pipeline gas, taking into consideration the evolving particularities of LNG (both commercial and technical) and the uncertainties related to decarbonisation. This book focuses on the specific role of LNG in the global energy transition as the world moves to decelerate rates of climate change.

Outline of the book

To date, LNG has been mainly treated as a method of transporting natural gas rather than an entity worthy of its own study. This book aims to fill that gap by studying LNG in detail and evaluating the differences between LNG and pipeline gas in a world in transition, taking into consideration the evolving commercial and technical characteristics of LNG and the uncertainties related to decarbonisation. An understanding of LNG's current place and potential in the energy market requires an analysis of the institutional, political, and resource characteristics that affect decision-making. I introduce a comprehensive framework to enable detailed analysis and integration of each of these aspects, with the ultimate aim of furthering our understanding of energy decisions, particularly in relation to LNG.

This book thus analyses the role of LNG in a decarbonising world from the perspective of LNG importers and the industry itself. Major investments in technology and know-how have occurred in recent years in an attempt to meet global demand, but energy systems require radical new pathways to meet climate goals in line with the Paris Agreement (UNFCCC, 2015). I set out to explore the role of LNG in the context of the global energy transition, arguing that LNG has a role to play that is influenced by four interconnected factors: (1) resources, (2) the LNG market, (3) energy-transition dynamics/regime status, and (4) geopolitical powers. These four factors are combined into a framework presented in Chapter 2 to give a broad overview of the current state of LNG in the global energy transition. Given the complex institutional, political, and resource characteristics that influence energy use, I introduce the T-E-T-I theoretical framework (combining techno-economic, transition, and institutional aspects; Al-Kuwari, 2023) to help make sense of complex scenarios involving energy use and transitions.

"Energy transition" is a general term that defines the current phase of our energy system evolution. However, regions and countries transition differently based on domestic and international objectives and constraints, including the availability of domestic resources, economic agendas, social pressures, and political systems. Chapters 3–5 focus specifically on the potential role of LNG in a decarbonising world through the in-depth analysis of three major energy markets – the UK (Chapter 3), Japan (Chapter 4), and the People's Republic of China (Chapter 5). These countries have different decarbonisation strategies, resource profiles, political systems, and economic forecasts, thus providing abundant material for in-depth study. The analyses identify potential pathways for development in the different energy systems, analysed from both an industry and a country perspective based on literature and in-depth interviews with stakeholders in the industry in each of the three countries (based on Al-Kuwari, 2023). Readers interested in the full methodology and the complete interview transcripts should consult Al-Kuwari (2023) for further information on the original study on which this text is based.

The UK was the first country to transition to fossil fuels and is now a global leader in decarbonisation and climate policy development. Historically a major natural gas and coal exporter, it has positioned itself for flexibility with multiple LNG and pipeline gas connections and an increased policy focus on renewable energy and hydrogen fuel development. Although the UK has aggressive decarbonisation targets, it faces significant challenges in hard-to-abate sectors such as heating. A particular focus of Chapter 3 is the role that LNG could play in the context of an energy system that is currently heavily dependent on natural gas for security of supply and seasonal demand, but that may undergo rapid decarbonisation.

In contrast, Japan (Chapter 4) imports fossil fuel to meet nearly all of its energy requirements and has limited renewable energy potential. The 2013 Fukushima disaster and long-standing ambitious climate and decarbonisation goals have placed the future of nuclear energy in doubt, and Japanese policymakers must manage competing objectives during the country's low-carbon energy transition. Japan was an LNG pioneer, and imports have increased recently to meet supply shortfalls. However, in the medium to long term, the role of LNG will be affected by the Government's energy policies. The world's first and largest LNG market is therefore currently uncertain about whether it will meet its 2050 hydrogen and energy efficiency targets. The focus of Chapter 4 is thus the impact of Japan's lack of pipeline gas access and how that determines the role of LNG, and, in the longer term, the potential impact of hydrogen fuel development on LNG exporters.

Chapter 5, the third and final case study, considers China's dramatic economic development over the past three decades. Although domestic coal resources are significant in that country, social pressure to improve air quality raises significant doubts about the continued role of coal in the energy system. The use of natural gas has grown significantly in recent years, supported by fuel-switching policies. LNG exporters will have to contend with pipeline gas-on-gas competition and increased proportions of nuclear and renewable fuels in the energy system. Chapter 5 thus evaluates the role of LNG in the context of price affordability, China's air-quality concerns, and a competitive energy environment.

Chapter 6 returns to a more macro-perspective, focusing broadly on the global future of LNG and how the industry could address decarbonisation from technological and commercial perspectives. Predictions about natural gas are made in the historic context of rapid and unpredictable changes. Again, I draw on interviews with policymakers, social scientists, industry leaders, and regulators to reflect political, economic, social, and institutional factors (Al-Kuwari, 2023). The chapter explores implications for LNG exporters in the context of the energy transition, comparing the characteristics of LNG with pipeline gas. This ties into a broader opportunity to identify potential pathways for LNG beyond 2030. Given that LNG supplies are rapidly increasing, exporters must consider how the risk of unused or stranded assets might be reduced by future-proofing either parts of or the entire

LNG chain. A growing body of grey literature has identified different pathways for the short- and medium-term energy transition of LNG, but a gap remains in terms of understanding whether LNG has viable medium- and long-term pathways compared with other solutions (particularly those focused on the hydrogen industry), and understanding the implications of these issues for exporters.

Climate change

The United Nations Framework Convention on Climate Change (UNFCCC) acknowledged that human activity has led to an increase in atmospheric concentrations of GHGs and warned that "additional warming of the Earth's surface and atmosphere may adversely affect natural ecosystems and humankind" (UNFCCC, 1992, p. 1). The convention represented a major milestone by laying out commitments from 155 signatory countries, including stabilising GHG emissions over the long term and limiting the emissions of harmful gases by developed countries based on targets and timetables, and establishing a GHG inventory. The Kyoto and Paris Agreements were two further milestones in climate discussions (UNFCCC, 1997, 2015). The Kyoto Protocol marked the first occasion on which binding GHG emission commitments were made by developed countries. The Paris Agreement saw 195 UNFCCC member countries agreeing to hold the increase in the global average temperature to below 2°C relative to pre-industrial levels and limit the temperature increase to 1.5°C above pre-industrial levels (UNFCCC, 2015, p. 3). Countries must report their progress towards these goals through nationally determined contributions (NDCs) every five years (McCauley & Heffron, 2018; Rogelj et al., 2016).

The IPCC was launched in 1988 by the World Meteorological Organization and the UN Environment Programme, and endorsed by the UN General Assembly (IPCC, 2001) to provide comprehensive assessments relevant to understanding climate change. The first report predicted an "average rate of increase of global mean temperature during the next century of about 0.3°C per decade with an uncertainty range from 0.2°C to 0.5°C per decade" (IPCC, 1990, p. 52). Subsequent reports provided more detail on the impacts of climate change, and adaptation and mitigation strategies. In 2015, the IPCC issued the Fifth Assessment Report (IPCC, 2014) (leading to the Paris Agreement), which concluded that

> without additional mitigation efforts beyond those in place today, and even with adaptation, warming by the end of the 21st century will lead to high to very high risk of severe, widespread and irreversible impacts globally.
>
> (p. 17)

It was anticipated that the trajectory at the time would increase global mean temperatures by 3.7–4.8°C in 2100 relative to pre-industrial levels (IPCC, 2014). The IPCC Special Report on Global Warming of 1.5°C (IPCC, 2018b) gave an updated assessment of the impacts of a global increase of temperature of 1.5°C and 2°C, concluding that holding temperature rises to 1.5°C would avoid the worst effects of a 2°C scenario. However, that increase would still have major consequences, including large-scale drought, loss of ecosystems, and an increase in the number of people living in poverty. The report emphasised the concept of a remaining carbon budget, defined as the

> estimated cumulative net global anthropogenic CO_2 [carbon dioxide] emissions from a given start date to the time that anthropogenic CO_2 emissions reach net zero that would result (at some probability) in limiting global warming to a given level, accounting for the impact of other anthropogenic emissions.
>
> (IPCC, 2018a, p. 24)

The report estimated that between 420 gigatons of carbon dioxide (Gt CO_2) and 580 Gt CO_2 remain, with a 50–66% probability of limiting warming to 1.5°C with current emissions at approximately 42 Gt CO_2 per year. Researchers continue to debate this number, which is based on wide-ranging assumptions for estimating natural- and anthropogenic-origin GHG increases, particularly methane (CH_4). The predictions also depend on policy decisions – for instance, in relation to fossil fuel production (Rogelj et al., 2019). The implications of the carbon budget are significant for policymakers, particularly for countries that depend on fossil fuels for their economic development.

Several oil- and gas-producing nations are in a paradoxical position in relation to climate change mitigation. Geography and climate mean that they will be adversely affected, but revenues from fossil fuels are well positioned to adapt in the short to medium term (Dell et al., 2008; Pal & Eltahir, 2015). Reduced future revenues from fossil fuel exports could dramatically affect the ability of existing and potential future producers to adapt to climate change. This paradigm – at the heart of the Paris Agreement – touches on how climate change adaptation should be funded, taking into consideration fossil fuel development and its role in supporting economies. McGlade and Ekins (2015) argued that GHG emissions for current estimates of fossil fuel reserves represent approximately three times more than what can be utilised based on the cumulative carbon emissions compatible with a warming of 2°C until 2050. This leads to the likelihood of being left with substantial amounts of un-burnable carbon – known fossil fuel reserves that should remain unburned because of climate change mitigation policies (Budinis et al., 2016; Pellegrini et al., 2024; Welsby et al., 2021). Countries that depend on fossil fuel exports for their development – or have plans to do so – are therefore directly affected by global climate policies and the associated energy transition (Al-Kuwari, 2023).

The negative impact of cumulative GHG emissions on our climate is conclusive despite the lack of clarity over the scale of the environmental and economic impact. Warming caused by CO_2 is irreversible over centuries, unless active measures are taken to remove the gas from the atmosphere (Al-Yafei et al., 2021; IPCC, 2014). These changes in temperature are projected to affect biodiversity, leading to species extinction (Pigot et al., 2023; Wabnitz et al., 2018) and causing parts of the world to become uninhabitable (IPCC, 2014; Pal & Eltahir, 2015). The effect will be disproportionate in different regions. Countries located in historically warmer areas of the world will feel the impact of climate change more keenly – this is particularly relevant because many developing countries are located in the warmer climates of the global south (McCauley & Heffron, 2018). The IPCC (2014, p. 54) reported a general consensus among researchers that "people who are socially, economically, culturally, politically, institutionally or otherwise marginalised are especially vulnerable to climate change and also to some adaptation and mitigation responses".

The energy transition

The world's energy mix is dominated by fossil fuels, which represent more than 80% of the global primary energy demand, followed by hydropower and renewable energy (nearly 14%), and nuclear energy (approximately 4%) (Al-Yafei et al., 2021; Energy Institute, 2024; IEA, 2018b). Oil dominates the energy sector in meeting demand for transportation and chemicals, and, historically, coal was a major player in electricity generation and industry. In recent decades, however, natural gas has competed directly with coal in both sectors and come to play a major role in heating in many regions (BP, 2019a; ExxonMobil, 2019; IEA, 2018b, 2021). Renewables have also been growing at a substantial rate in the power sector, becoming a major source of competition for both coal and natural gas (BP, 2022; IEA, 2018b; IRENA, 2018). Recent trends indicate that the demand for coal is likely to peak within the next few years, together with an increase in demand for natural gas and oil, while nuclear energy demand has stagnated in most regions (BP, 2019a; IEA, 2024). In 2018, renewables represented less than 5% of the energy mix (IEA, 2018b). However, their use is now growing rapidly on account of dramatic cost reductions (BP, 2022; Energy Institute, 2024), and in 2023, renewables represented almost 15% of global primary energy consumption, and a further significant growth is anticipated in the next five years (BP, 2022; Energy Institute, 2024; IEA, 2024, n.d.). The share of oil in the energy supply has remained stable, although the historic volatility of oil prices has encouraged importing-country governments to promote policies to decrease oil dependence (Selden, 2020; Van de Graaf et al., 2016). Coal is abundant and more affordable than natural gas in many parts of the world (Stern, 2017), but its environmental characteristics have put pressure on future adoption, while natural gas has grown steadily in almost all sectors and regions (IEA, 2016, 2018b; Kim et al., 2023). Figures 1.1 and 1.2 summarise the evolution of global primary energy consumption. The decline in demand in 2020 was linked to the COVID-19 pandemic.

Oil, coal, and natural gas have different characteristics, grades, and qualities that affect air quality. Oil is consumed largely in the form of products from refineries that are used in mobile applications, mostly transportation, and its use produces sulfur dioxide (SO_2) and nitrogen oxides (NO_X). Coal contains impurities such as metals and sulfur and thus requires treatment to minimise the adverse impacts of these. It is responsible for approximately 60% of the global consumption of SO_2, which can cause respiratory illness. Power generation is a major source of SO_2, with coal being the main fuel responsible for air pollution. Most inhalable

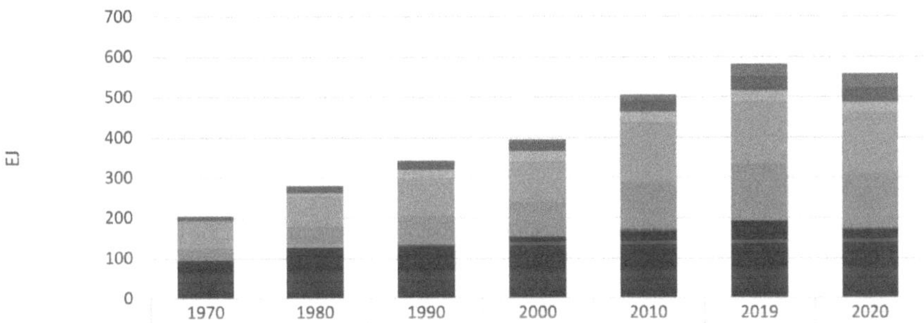

	1970	1980	1990	2000	2010	2019	2020
Renewables	0.3	0.6	1.57	2.64	9.63	28.82	31.71
Hydroelectric	11.75	17.32	21.59	26.52	32.25	37.69	38.16
Nuclear Energy	0.79	7.12	20	25.81	25.99	24.93	23.98
Coal	61.41	75.09	93.23	98.73	151.21	157.64	151.42
Natural gas	34.61	51.26	70.14	86.43	113.78	140.54	137.62
Oil	95.34	128.02	135.6	154.39	172.53	191.89	174.2

Figure 1.1 Historical global primary energy demand (BP Statistical Review of World Energy, 2021)

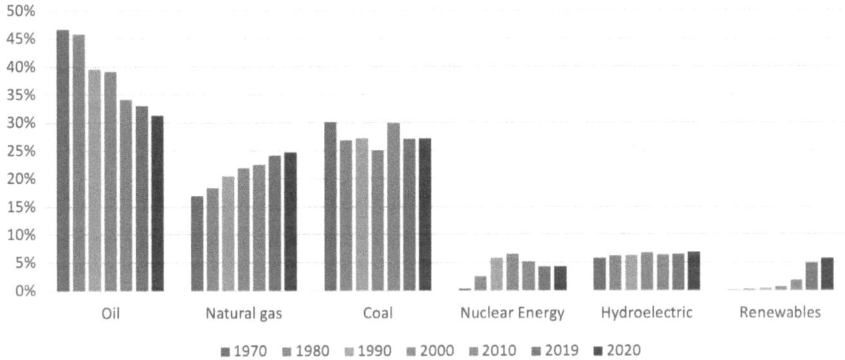

Figure 1.2 Historical global primary energy demand by percentage and fuel (BP Statistical Review of World Energy, 2021)

particulate matter ($PM_{2.5}$) emissions come from buildings via cooking, heating, and lighting, while industry represents nearly half of SO_2 emissions, and the transportation sector is responsible for nearly half of NO_X emissions. From an air-quality perspective, nearly all sulfur oxides (SO_X) and most NO_X emissions can be eliminated by replacing a coal-generating unit with an efficient gas-fired combined-cycle plant (Chandra, 2017). Natural gas has the lowest volumes of emissions, partly owing to the lack of sulfur, which is removed during processing to limit pipeline damage from corrosion (Al-Yafei et al., 2021; IEA, 2016).

Concerns over air quality drive global fuel-switching policies, which have initially focused on coal to gas in power generation and industry (Qin et al., 2018;

Wang et al., 2023). In some cases, improved economics have allowed countries to move straight from coal to renewables, with gas being the focus in more industrial processes, as evidenced by China's successful clear air strategy (Qin et al., 2018). Increased air pollution controls and electrification, coupled with renewable sources of energy, have helped to accelerate the decoupling of fuel combustion and pollution and positively affected air quality.

From strategic and political perspectives, the use of coal and gas introduces different risks. Coal is typically sourced domestically and is flexible in terms of storage and transportation, whereas gas lacks flexibility if delivered by pipeline and generally requires more expensive infrastructure. Politically, the coal industry has strong support in producer countries such as China, Germany, and South Africa – where it is typically a major employer (Colombo et al., 2016; Zhang et al., 2023). Economically, gas will increasingly have to compete with renewables for power generation, as the costs of solar and wind power production continue to decline (Osman et al., 2023). Furthermore, despite potential synergies that flexible gas-fired power plants provide to support the intermittent nature of renewables, the case for switching is somewhat undermined by battery storage technology and the additional efficiency of newly commissioned coal-fired plants (Boersma & Jordaan, 2017). Compared with imported natural gas, renewables provide advantages related to foreign exchange costs, domestic employment, and foreign dependence on energy (Boersma & Jordaan, 2017; Chandra, 2017; IEA, 2018a, 2018b; Moniz et al., 2011; Stern, 2017). There are some opportunities for switching from oil to gas for chemical and industrial uses (ETC, 2018) and within the maritime industry (IEA, 2016). However, the ease of transport and storage, and the higher energy density of oil and more fragmented and complicated business models of natural gas present challenges to the practicality of this approach (Al-Yafei et al., 2021; Stern, 2017).

Globally, fuel-switching and climate policies are a key driver in future energy demand. The IPCC (2014, p. 32) concluded:

> Greenhouse gas emissions from energy supply can be reduced significantly by replacing current world average coal-fired power plants with modern, highly efficient natural gas combined-cycle power plants or combined heat and power plants, provided that natural gas is available, and that the fugitive emissions associated with the extraction and supply of natural gas are low or mitigated.

Natural gas can lower overall emissions subject to managing CH_4 emissions (IEA, 2018b; IEA, 2020). However, countries must implement policies to substitute coal with gas to ensure that there is no incremental supply of fossil fuels to the energy system (Al-Yafei et al., 2021; Gilbert & Sovacool, 2017; Lazarus et al., 2015). A study by the National Energy Technology Laboratory (Roman-White et al., 2019) concluded that over a 100-year period, power generated in Europe and Asia from exported natural gas from a variety of sources has lower life-cycle GHG emissions than that from regionally sourced coal-power generation. Countries can achieve rapid emission reductions by switching to natural gas and consequently benefiting from air-quality improvements (Bayomi et al., 2017; Qin et al., 2018).

Natural gas

Natural gas is a fossil fuel containing CH_4 (65–95%), ethane (2–15%), propane (0.2–5%), butane (0–5%), and pentane (0.05–2%) (Chandra, 2017). Natural gas liquids such as propane and butane may also be present depending on the characteristics of the reservoir – all are fuels with their own value and use in industrial and residential markets. Natural gas contains non-hydrocarbon components, some of which are impurities, such as water, and others of which are more environmentally damaging components, such as CO_2, hydrogen sulfide (H_2S), and helium. Natural gas occurs in associated and non-associated gas fields – reservoirs that coexist with crude oil and ones that comprise only natural gas, respectively. Methane produces lower CO_2 emissions than other fossil fuels when combusted, although CH_4 itself is more potent (Smil, 2015).

The gaseous form of natural gas has a significantly lower energy density, making it a more challenging fuel to transport and store. Only when the gas is liquified to –162°C, delivering a 600-fold improvement in density, does it reach similar volumetric density to wood or other liquid fuels (Kim et al., 2023; Smil, 2015). Once extracted, natural gas requires processing and treatment to recover liquids and remove non-toxic elements, both of which affect project economics by providing additional revenue streams (i.e. ethane and propane) and maximising the amount of available energy (Chandra, 2017). In addition, environmentally damaging and corrosive components that can cause air pollution contribute to global warming, and physically corroded pipelines must be removed. CO_2 and H_2S, which can cause emissions of SO_X, are the principal components to be removed before the gas is sold. CO_2 disposal presents a challenge for gas producers because they must factor in the economic impact and technical considerations and navigate local and international environmental regulations. However, natural gas processing is essentially a process of separation and removal of elements to meet environmental and quality standards, and, thus, it is generally a simpler process than for crude oil, where fuel refining and production require complex processes involving high temperatures and pressures (Chandra, 2017; Moniz et al., 2011). Reservoir profiles have a significant impact on development decisions because the costs of removing the harmful components must be weighed against the potential benefits of selling the natural gas and any associated liquids.

After processing, gas has historically been shipped via pipelines, with the first high-pressure lines operating in the United States in the late nineteenth century (Chandra, 2017). The volume of gas transported depends on pipeline diameter and operating pressure, typically supported by compressors along the line. Steel is the main component of pipelines and accounts for approximately 45% of the cost of large-scale projects. Pipelines are designed to deliver natural gas to consumers, including power plants, industrial facilities, and homes. Gas storage also plays a role in many regions to meet long- and short-term requirements. Depleted gas reservoirs, aquifers, and salt caverns are three types of natural underground storage options, each of which has its own characteristics and economics (Moniz et al., 2011). Most of the world's gas-storage reservoirs are located in Europe and the United States, and combined with over-ground LNG storage, provide much-needed flexibility to meet seasonal demand, such as high heating requirements in the northern hemisphere during winter and high air-conditioning demand in the Middle East in the summer (Chandra, 2017; Dahan & Altheeb, 2023). By contrast, crude oil pipeline transportation and storage is more economical because liquids are easier to transport (Moniz et al., 2011).

LNG is another form of natural gas transportation, allowing for more flexible and dynamic delivery to global markets. Over the past decade, most of the growth in natural gas supply has been through LNG – an indication of its technical maturity and strategic and commercial benefits (Al-Kuwari, 2023). Natural gas emits 30% less CO_2 per kilowatt hour than oil, and 45% less CO_2 than coal when combusted (Al-Yafei et al., 2021; Chandra, 2017; National Grid, 2024). Conversely, CH_4, as well as being a fossil fuel, is a potent GHG that traps heat more than 20 times more effectively than CO_2, although its atmospheric residence time is only a fraction of that of CO_2 (Anderson & Broderick, 2017; Chandra, 2017; Smil, 2015). CH_4 is responsible for nearly one-fifth of the increase in anthropogenic radiative forcing in the past 250 years (Smil, 2015) and continuous high CH_4 emissions would maintain this warming effect (Anderson & Broderick, 2017). Since 2015, CH_4 emissions from the oil and gas sector have increased by an average of 2% per year, with 29 megatons (Mt) of the 79 Mt in 2017 coming from natural gas, and 80% overall coming from upstream processes (IEA, 2018b). Natural gas leaks from infrastructure are, thus, a long-term challenge for the industry (Gilbert & Sovacool, 2017; IEA, 2023; Tavoni, 2015). Despite the negative impact of CH_4, CO_2 emissions have a greater long-term temperature change effect owing to the total quantity of natural gas production and the pollutant's persistence in the atmosphere. Thus, there is a need to focus on decarbonisation and CH_4 emission abatement.

Most studies describe a time-defined period for the role of natural gas, either by stating a peak-and-plateau period or a limit for use as a fuel in energy systems, all depending on the level of climate goal ambition. Levi (2013) and Zhang et al. (2016) concluded that natural gas can play a role if a more modest and longer bridging period is anticipated. Others demonstrated that natural gas demand needs to peak within the next two to three decades, before being substituted by lower-carbon energy sources, mainly renewables (Al-Yafei et al., 2021; Holz et al., 2015; McGlade & Ekins, 2014; McGlade et al., 2018). The studies vary, however, in how rapid that substitution process needs to be. Limiting the environmental impact of natural gas production by reducing CH_4 emissions from flaring, venting, and leaking is another precondition for maintaining the position of LNG as an environmentally preferable option over other fossil fuels (Anderson & Broderick, 2017; Erickson et al., 2015; Zhang et al., 2016). Natural gas substituting for coal in energy systems should be introduced alongside policies that discourage demand growth for fossil fuels to reduce the risk of carbon lock-in and other infrastructure development that pushes out renewables and delays decarbonisation (Hausfather, 2015; Lazarus et al., 2015; Littell, 2017; McGlade et al., 2018; Shearer et al.,

2014). McGlade and Ekins (2014) concluded that natural gas must be substituted for coal, and the increase of gas must be supported by an even larger decrease in coal consumption to meet a global temperature target in line with climate obligations. Carbon capture and storage (CCS) also needs to be available and deployable at scale for gas to play a role as a transition fuel and gas decarbonisation tool (McGlade et al., 2018; Stern, 2019; Zhang et al., 2016).

Liquified natural gas

LNG is a liquid product made by cooling natural gas to approximately –162°C, in the process reducing its volume 600-fold compared with its gaseous state (Kim et al., 2023; Kumar et al., 2011). The liquid can then be stored and transported by vessel to markets across the world, making it a viable alternative for transporting large volumes of gas. The LNG chain consists of several components (discussed in Chapter 2). The extraction (or upstream) process is the same as for the gaseous product. However, once the piped gas is transported to the liquefaction terminal, it is transformed into liquid form (Bridge & Bradshaw, 2017). Impurities are removed before cooling to avoid damaging the plant's equipment. After liquefaction, specially designed vessels transport the LNG to international receiving regasification terminals. Here, the LNG is converted back into gas and transported via pipeline to local consumers (Kim et al., 2023). Throughout the process, the gas remains chemically unchanged.

LNG projects are complex, risky, and expensive (Lazarus et al., 2015; Tusiani & Shearer, 2016). The processes involved in successful LNG delivery require significant gas reserves, deep-water ports, access to capital, long-term commitment from buyers, and transparent government regulations. Despite an increase in LNG projects, overall costs remain challenging for the industry (Al-Yafei et al., 2021; Kim et al., 2023; Smil, 2015; Stern, 2017). The liquefaction step is typically the most expensive because of the sophisticated technology needed, high operating costs, and large energy requirements, representing a significant proportion of the capital costs, followed by shipping and regasification (Chandra, 2017; Kim et al., 2023; Tusiani & Shearer, 2016). In the past decade, costs have increased at a higher rate than for oil and pipeline gas owing to the complexity of new projects and increased demand on a handful of established technology providers and engineering firms (Chandra, 2017). Corbeau and Ledesma (2016) estimated that the associated costs of upstream processing and liquefaction may be double that of oil, and three to four times more than coal in an equivalent cost on an energy basis – a 5-Mt-per-annum LNG plant produces the equivalent of a 120,000 billion barrels of oil per day or 10.8 Mt of coal. Despite the high costs, LNG provides an opportunity for previously (or partly) stranded gas to be delivered to new markets and revenue streams.

Decarbonisation

Decarbonisation is defined as delivering affordable energy services with fewer GHG emissions (Bridge et al., 2013). The decarbonisation of fossil fuels in energy systems can occur either by decarbonising the fuels prior to burning or by capturing the emitted CO_2 (Abbasi & Abbasi, 2011). To meet the Paris Agreement's climate change mitigation plans (UNFCCC, 2015), significant technical and policy-driven decarbonisation strategies are being pursued by managing short-term energy demand to alter our energy system structure (Al-Yafei et al., 2021; Ekins, 2010; Grübler et al., 1999). In the short term, low-carbon energy sources must increase, while global energy system efficiencies improve (IPCC, 2014). In the long term, energy systems must move towards deep decarbonisation. Some measures suggest that decarbonisation has already taken place. Aguilera and Aguilera (2012) demonstrated that by analysing the hydrogen-to-carbon ratio of global energy systems, with hydrogen as a proxy for environmental quality, the percentage of hydrogen in energy systems increased from 1850 to 1970. This was due to the introduction and use of fuels with progressively less carbon – from coal to oil to natural gas. Despite the improved rate of decarbonisation during that time, the total volume of emissions increased substantially (Grübler & Nakićenović, 1996).

Electricity from renewable and nuclear sources provides a clear path for decarbonisation (Staffell et al., 2018, IEA, 2023). However, hard-to-abate sectors, which account for approximately 30% of global emissions, are currently forecast to increase and present a challenge to achieving full decarbonisation (BP, 2019a, 2019b; ETC, 2018; IEA, 2018b; Royal Dutch Shell, 2018; Sharmina et al., 2020). The sectors are broadly categorised into two groups – heavy industry (mainly cement, steel, and chemicals production) and heavy-duty transport (e.g. shipping and aviation). In countries such as the UK and Canada, heating is also a significant challenge because it represents complicated seasonal demand with high strategic importance (Staffell et al., 2018). Hard-to-abate sectors are difficult to address, either because emission reduction costs are too high or because alternative solutions and technologies are insufficiently mature, such as efficient large-scale electric batteries (IEA, 2018b). The creation of new materials, cheaper electrolysis to support more economical hydrogen-production methods – a key alternative to natural gas – and carbon capture are also areas where innovation is needed. The ETC (2018) defined a specific path to address the harder-to-abate sectors and concluded that it would be technically possible to achieve net-zero at a cost of less than 0.5% of global gross domestic product (GDP). Edenhofer et al. (2009) used a series of global energy–environment–economy models to demonstrate that stabilising CO_2 emissions can range from 0.8% to 2.5% of global GDP. Despite criticism (Nordhaus, 2007), Lord Stern's study on the economics of climate change (Stern, 2006, p. 188) concluded that if no mitigative action is taken, "the cost of BAU (business as usual) climate change over the next two centuries [will be] equivalent to a loss of at least 5% of global per-capita consumption, now and forever". Others have made the case that a low-carbon transition can bring economic value. The International Renewable Energy Agency (IRENA) (2018), for instance, demonstrated that investment into energy transition would bring relative improvements to global GDP and in some scenarios could pay for itself in cumulative terms up to US$160 trillion until 2050.

Despite being characterised as a more climate advantageous fossil fuel on account of its efficient combustion characteristics, natural gas contains impurities, mainly CO_2 (Boersma & Jordaan, 2017; IPCC, 2011; Lazarus et al., 2015; Tavoni, 2015). Although harmful pollutants, including SO_2, NO_X, and particulate matter, contribute directly to poor air quality and pollution (IEA, 2016; Myhrvold & Caldeira, 2012; Pierce, 2011; Stephenson et al., 2011), the combustion of CH_4 generates less CO_2 per unit than coal, wood, or crude or refined oil products (Smil, 2015). Overall, however, natural gas has a relatively higher hydrogen-to-carbon ratio, and relatively lower emissions of contaminants (Abánadcs, 2018; Chandra, 2017; IEA, 2016; IMF, 2014; IPCC, 2014; Qin et al., 2018).

GHG emission projections vary widely based on climate policy and socio-economic development, with most forecasts being incompatible with 2°C scenarios. Continuing population growth and increasing prosperity in the developing world are forecast to continue to put pressure on carbon emissions (Al-Yafei et al., 2021; BP, 2019a; ExxonMobil, 2019; Tamburino et al., 2023). Although there will be a reduction in carbon intensity because of the changing fuel mix and rapid growth of electricity, it will largely be offset by increases in overall electricity consumption (BP, 2019a; ExxonMobil, 2019). In the IEA's (2018b) Stated Policies Scenario (STEPS), based on the current policy landscape laid out following the Paris Agreement (UNFCCC, 2015), energy-related CO_2 emissions trend gradually upwards to 2030 (IEA, 2021), with gas demand in 2030 higher than today, while Moniz et al. (2011, p. 2) tout natural gas as a "cost effective bridge to a low-carbon future". Thus, although natural gas is not the only answer to mitigate the negative impacts of climate change, it does allow more time to develop strategies and alternatives to move towards a low-carbon future, while contributing to improved air quality. However, a clear set of preconditions is needed, focused on timing, climate ambition, CH_4 emissions, substitution, carbon lock-in, and the availability of CCS technology.

References

Abánades, A. (2018). Natural gas decarbonization as tool for greenhouse gases emission control. *Frontiers in Energy Research*, *6*, 1–7. https://doi.org/10.3389/fenrg.2018.00047

Abbasi, T., & Abbasi, S. A. (2011). Decarbonization of fossil fuels as a strategy to control global warming. *Renewable and Sustainable Energy Reviews*, *15*(4), 1828–1834. https://doi.org/10.1016/j.rser.2010.11.049

Aguilera, R. F., & Aguilera, R. (2012). World natural gas endowment as a bridge towards zero carbon emissions. *Technological Forecasting and Social Change*, *79*(3), 579–586. https://doi.org/10.1016/j.techfore.2011.09.004

Al-Kuwari, O. (2023). *The future of liquified natural gas (LNG) in the energy transition: Options and implications for the LNG industry in a decarbonising world* [Unpublished doctoral dissertation]. University College London. https://discovery.ucl.ac.uk/id/eprint/10173650/

Al-Yafei, H., Aseel, S., Kucukvar, M., Onat, N. C., Al-Sulaiti, A., & Al-Hajri, A. (2021). A systematic review for sustainability of global liquified natural gas industry: A 10-year update. *Energy Strategy Reviews*, *38*, 100768. https://doi.org/10.1016/j.esr.2021.100768

Anderson, K., & Broderick, J. (2017). *Natural gas and climate change*. University of Manchester. https://research.manchester.ac.uk/en/publications/natural-gas-and-climate-change

Bayomi, N., Heinrich, M., & Fernandez, J. E. (2017). Emission saving potentials from transition to natural gas usage: A view from four major emitters. *Energy Procedia*, *142*, 3147–3153. https://doi.org/10.1016/j.egypro.2017.12.377

Boersma, T., & Jordaan, S. M. (2017). Whatever happened to the golden age of natural gas? *Energy Transitions*, *1*(2), 1–8. https://doi.org/10.1007/s41825-017-0005-4

BP (British Petroleum). (2019a). *BP energy outlook 2019*. www.bp.com/content/dam/bp/business-sites/en/global/corporate/pdfs/energy-economics/energy-outlook/bp-energy-outlook-2019.pdf

BP (British Petroleum). (2019b). *BP statistical review of world energy 2019*. www.bp.com/content/dam/bp/business-sites/en/global/corporate/pdfs/energy-economics/statistical-review/bp-stats-review-2019-full-report.pdf

BP (British Petroleum). (2021). *BP statistical review of world energy 2021*. www.bp.com/content/dam/bp/business-sites/en/global/corporate/pdfs/energy-economics/statistical-review/bp-stats-review-2021-full-report.pdf

BP (British Petroleum). (2022). *BP energy outlook 2022*. www.bp.com/content/dam/bp/business-sites/en/global/corporate/pdfs/energy-economics/energy-outlook/bp-energy-outlook-2022.pdf

Bridge, G., Bouzarovski, S., Bradshaw, M., & Eyre, N. (2013). Geographies of energy transition: Space, place and the low-carbon economy. *Energy Policy*, *53*, 331–340. https://doi.org/10.1016/j.enpol.2012.10.066

Bridge, G., & Bradshaw, M. (2017). Making a global gas market: Territoriality and production networks in liquefied natural gas. *Economic Geography*, *93*(3), 215–240. https://doi.org/10.1080/00130095.2017.1283212

Budinis, S., Krevor, S., MacDowell, N., Brandon, N., & Hawkes, A. (2016). *Can technology unlock 'unburnable carbon'?* Sustainable Gas Institute, Imperial College London. www.imperial.ac.uk/media/imperial-college/research-centres-and-groups/cpse/annual-report/newsletter-and-technical-articles/Technical-Article – Newsletter-Issue–15–July-2017.pdf

Chandra, V. (2017). *Fundamentals of natural gas: An international perspective*. PennWell Books.

Colombo, S., El Harrak, M., & Sartori, N. (Eds.). (2016). *The future of natural gas: Markets and geopolitics*. European Energy Review. https://doi.org/10.1016/0040-1625(87)90047-3

Corbeau, A., & Ledesma, D. (Eds.). (2016). *LNG markets in transition: The great reconfiguration*. Oxford University Press for the Oxford Institute for Energy Studies and the King Abdullah Petroleum Studies and Research Center.

Dahan, A., & Altheeb, R. (2023). Can the Arab's natural gas secure the Europeans' gas requirements? The case of liquified natural gas (LNG). *Uncertain Supply Chain Management*, *11*(4), 1677–1684. https://doi.org/10.5267/j.uscm.2023.7. 001

Dell, M., Jones, B., & Olken, B. (2008). *Climate change and economic growth: Evidence from the last half century*. National Bureau of Economic Research. https://doi.org/10.3386/w14132

Edenhofer, O., Knopf, B., Barker, T., Baumstark, L., Kitous, A., Kypreos, S., Leimbach, M., Magne, B., Scrieciu, S., & Turton, H. (2009). Low stabilization pathways: Economic and technical feasibility. *IOP Conference Series: Earth and Environmental Science*, *6*(27), 272002. https://doi.org/10.1088/1755-1307/6/7/272002

Ekins, P. (Ed.). (2010). *Hydrogen energy: Economic and social challenges*. Earthscan.

Energy Institute. (2024). *Statistical review of world energy* (73rd ed.). www.energyinst.org/__data/assets/pdf_file/0006/1542714/EI_Stats_Review_2024.pdf

Erickson, P., Kartha, S., Lazarus, M., & Tempest, K. (2015). Assessing carbon lock-in. *Environmental Research Letters*, *10*(8). https://doi.org/10.1088/1748-9326/10/8/084023

ETC (Energy Transitions Commission). (2018). *Mission possible: Reaching net-zero carbon emissions from harder-to-abate sectors by mid-century*. www.energy-transitions.org/publications/mission-possible/

ExxonMobil. (2019). *Outlook for energy: A perspective to 2040*. www.ief.org/_resources/files/events/ief-lecture-exxonmobils-2019-outlook-for-energy-a-perspective-to-2040/exxonmobil-2019-outlook-for-energy-a-perspective-to-2040.pdf

Gilbert, A. Q., & Sovacool, B. K. (2017). US liquefied natural gas (LNG) exports: Boom or bust for the global climate? *Energy*, *141*, 1671–1680. https://doi.org/10.1016/j.energy.2017.11.098

Grübler, A., & Nakićenović, N. (1996). Decarbonizing the global energy system. *Technological Forecasting and Social Change*, *53*(1), 97–110. https://doi.org/10.1016/0040-1625(96)00049-2

Grübler, A., Nakićenović, N., & Victor, D. G. (1999). Dynamics of energy technologies and global change. *Energy Policy*, *27*(5), 247–280. https://doi.org/10.1016/S0301-4215(98)00067-6

Hausfather, Z. (2015). Bounding the climate viability of natural gas as a bridge fuel to displace coal. *Energy Policy*, *86*, 286–294. https://doi.org/10.1016/j.enpol.2015.07.012

Holz, F., Richter, P. M., & Egging, R. (2015). A global perspective on the future of natural gas: Resources, trade, and climate constraints. *Review of Environmental Economics and Policy*, *9*(1). https://doi.org/10.1093/reep/reu016

IEA (International Energy Agency). (2016). *Energy and air pollution – world energy outlook: Special report 266*. IEA Publications. https://doi.org/10.1021/ac00256a010

IEA (International Energy Agency). (2018a). *Outlook for producer economies*. IEA Publications. www.iea.org/reports/outlook-for-producer-economies

IEA (International Energy Agency). (2018b). *World energy outlook 2018*. IEA Publications. www.iea.org/reports/world-energy-outlook-2018

IEA (International Energy Agency). (2020). *Methane tracker 2020*. IEA Publications. www.iea.org/reports/methane-tracker-2020

IEA (International Energy Agency). (2021). *Net zero by 2050 – A roadmap for the global energy sector*. IEA Publications. https://iea.blob.core.windows.net/assets/ad0d4830-bd7e-47b6-838c-40d115733c13/NetZeroby2050-ARoadmapfortheGlobalEnergySector.pdf

IEA (International Energy Agency). (2023). *World energy outlook 2023*. IEA Publications. www.iea.org/reports/world-energy-outlook-2023

IEA (International Energy Agency). (2024). *Renewables 2023: Analysis and forecast to 2028*. IEA Publications. www.iea.org/reports/renewables-2023

IEA (International Energy Agency). (n.d.). *World*. Retrieved July 16, 2024, from www.iea.org/world

IMF (International Monetary Fund). (2014). *World economic outlook October 2014: Legacies, clouds, uncertainties*. www.imf.org/en/Publications/WEO/Issues/2016/12/31/Legacies-Clouds-Uncertainties

IPCC (Intergovernmental Panel on Climate Change). (1990). IPCC 1990 and 1992 assessments. *Nature*, *360*(6402), 292. https://doi.org/10.1038/360292e0

IPCC (Intergovernmental Panel on Climate Change). (2001). *Climate change 2001: Synthesis report (third assessment report)*. www.un.org/en/climatechange/reports?gad_source=1&gclid=CjwKCAjwrIixBhBbEiwACEqDJXHRVdPbZzZJJhYEOIo06dSvzCo9dkIJzz7Ai1puNNIafErKCGMNORoCQgIQAvD_BwE

IPCC (Intergovernmental Panel on Climate Change). (2011). Anthropogenic and natural radiative forcing. In *Climate change 2013: The physical science basis. Contribution of working group I to the fifth assessment report of the intergovernmental panel on climate change* (pp. 23–56). Cambridge University Press.

IPCC (Intergovernmental Panel on Climate Change). (2014). *Climate change 2014: Synthesis report (assessment report five)*. www.ipcc.ch/report/ar5/syr/

IPCC (Intergovernmental Panel on Climate Change). (2018a). *IPCC SR15 full report*. https://archive.ipcc.ch/report/sr15/pdf/sr15_citation.pdf

IPCC (Intergovernmental Panel on Climate Change). (2018b). *Summary for policymakers: Global warming of 1.5 °C.* https://archive.ipcc.ch/report/sr15/pdf/sr15_citation.pdf

IRENA (International Renewable Energy Agency). (2018). *Hydrogen from renewable power: Technology outlook for the energy transition.* www.irena.org/publications/2018/sep/hydrogen-from-renewable-power

Kim, Y., Lee, J., An, N., & Kim, J. (2023). Advanced natural gas liquefaction and regasification processes: Liquefied natural gas supply chain with cryogenic carbon capture and storage. *Energy Conversion and Management*, *292*, 117349. https://doi.org/10.1016/j.enconman.2023.117349

Kumar, S., Kwon, H. T., Choi, K. H., Lim, W., Cho, J. H., Tak, K., & Moon, I. (2011). LNG: An eco-friendly cryogenic fuel for sustainable development. *Applied Energy*, *88*(12), 4264–4273. https://doi.org/10.1016/j.apenergy.2011.06.035

Lazarus, M., Tempest, K., Klevnäs, P., & Korsbakken, J. I. (2015). *Natural gas: Guardrails for a potential climate bridge.* The Global Commission on the Economy and Climate. https://policycommons.net/artifacts/10746308/natural-gase/11624063/

Levi, M. (2013). Climate consequences of natural gas as a bridge fuel. *Climatic Change*, *52*(6), 709–716. https://doi.org/10.1007/s10584-012-0658-3

Lin, N., & Brooks, R. E. (2021). Global liquified natural gas trade under energy transition. *Energies*, *14*(20), 6617. https://doi.org/10.3390/en14206617

Littell, D. (2017). Natural gas: Bridge or wall in transition to low-carbon economy? *Natural Gas Electricity*, *33*(6), 1–8. https://doi.org/1-8. 10.1002/gas.21953

McCauley, D., & Heffron, R. (2018). Just transition: Integrating climate, energy and environmental justice. *Energy Policy*, *119*, 1–7. https://doi.org/10.1016/j.enpol.2018.04.014

McGlade, C., & Ekins, P. (2014). Un-burnable oil: An examination of oil resource utilisation in a decarbonised energy system. *Energy Policy*, *64*, 102–112. https://doi.org/10.1016/j.enpol.2013.09.042

McGlade, C., & Ekins, P. (2015). The geographical distribution of fossil fuels unused when limiting global warming to 2 °C. *Nature*, *517*, 187–190. https://doi.org/10.1038/nature14016

McGlade, C., Pye, S., Ekins, P., Bradshaw, M., & Watson, J. (2018). The future role of natural gas in the UK: A bridge to nowhere? *Energy Policy*, *113*, 454–465. https://doi.org/10.1016/j.enpol.2017.11.022

Moniz, E., Jacoby, H., Meggs, A., Kenderdine, M., & O'Sullivan, F. (2011). *The future of natural gas – MIT study.* http://energy.mit.edu/wp-content/uploads/2011/06/MITEI-The-Future-of-Natural-Gas.pdf

Myhrvold, N. P., & Caldeira, K. (2012). Greenhouse gases, climate change and the transition from coal to low-carbon electricity. *Environmental Research Letters*, *7*(1). https://doi.org/10.1088/1748-9326/7/1/014019

National Grid. (2024, April 3). *What is liguefied natural gas?* National Grid Group. www.nationalgrid.com/stories/energy-explained/what-is-liquefied-natural-gas-lng

Nordhaus, W. D. (2007). A review of the Stern review on the economics of climate change. *Journal of Economic Literature*, *45*(3), 686–702. https://doi.org/10.1257/jel.45.3.686

Osman, A. I., Chen, L., Yang, M., Msigwa, G., Farghali, M., Fawzy, S., Rooney, D. W., & Yap, P.-S. (2023). Cost, environmental impact, and resilience of renewable energy under a changing climate: A review. *Environmental Chemistry Letters*, *21*, 741–764. https://doi.org/10.1007/s10311-022-01532-8

Pal, J. S., & Eltahir, E. A. B. (2015). Future temperature in southwest Asia projected to exceed a threshold for human adaptability. *Nature Climate Change*, *6*(2), 197–200. https://doi.org/10.1038/NCLIMATE2833

Pellegrini, L., Arsel, M., Muñoa, G., Rius-Taberner, G., Mena, C., & Orta-Martínez, M. (2024). The atlas of unburnable oil for supply-side climate policies. *Nature Communications*, *15*(2318). https://doi.org/10.1038/s41467-024-46340-6

Pierce, R. (2011). Natural gas: A long bridge to a promising destination. *Utah Environmental Law Review, 32*, 245. https://doi.org/10.3366/ajicl.2011.0005
Pigot, A. L., Merow, C., Wilson, A., & Trisos, C. H. (2023). Abrupt expansion of climate change risks for species globally. *Nature Ecology & Evolution, 7*(7), 1060–1071. https://doi.org/10.1038/s41559-023-02070-4
Qin, Y., Höglund-Isaksson, L., Byers, E., Feng, K., Wagner, F., Peng, W., & Mauzerall, D. L. (2018). Air quality carbon – water synergies and trade-offs in China's natural gas industry. *Nature Sustainability, 1*(9), 505–511. https://doi.org/10.1038/s41893-018-0136-7
Rogelj, J., Den Elzen, M., Höhne, N., Fransen, T., Fekete, H., Winkler, H., Schaeffer, R., Sha, F., Riahi, K., & Meinshausen, M. (2016). Paris Agreement climate proposals need a boost to keep warming well below 2 °C. *Nature, 534*(7609), 631–639. https://doi.org/10.1038/nature18307
Rogelj, J., Forster, P. M., Kriegler, E., Smith, C. J., & Séférian, R. (2019). Estimating and tracking the remaining carbon budget for stringent climate targets. *Nature, 571*, 335–342. https://doi.org/10.1038/s41586-019-1368-z3–10
Roman-White, S., Rai, S., Littlefield, J., Cooney, G., & Skone, T. J. (2019). *Life cycle greenhouse gas perspective on exporting liquefied natural gas from the United States: 2019 Update*. National Energy Technology Laboratory. www.osti.gov/biblio/1607677
Royal Dutch Shell. (2018). *Shell scenarios sky – meeting the goals of the Paris Agreement*. www.shell.com/news-and-insights/scenarios/what-are-the-previous-shell-scenarios.html
Selden, Z. (2020). Taming the bear: American liquified natural gas (LNG) exports and the negation of Russian influence in Europe. *Global Affairs, 6*(2), 149–165. https://doi.org/10.1080/23340460.2020.1780460
Sharmina, M., Edelenbosch, O. Y., Wilson, C., Freeman, R., Gernaat, D. E. H. J., Gilbert, P., Larkin, A., Littleton, E. W., Traut, M., Van Vuuren, D. P., Vaughan, N. E., Wood, F. R., & Le Quéré, C. (2020). Decarbonising the critical sectors of aviation, shipping, road freight and industry to limit warming to 1.5–2 °C. *Climate Policy, 21*(4), 455–474. https://doi.org/10.1080/14693062.2020.1831430
Shearer, C., Bistline, J., Inman, M., & Davis, S. J. (2014). The effect of natural gas supply on US renewable energy and CO2 emissions. *Environmental Research Letters, 9*(9), 1–9. https://doi.org/10.1088/1748-9326/9/9/094008
Smil, V. (2015). *Natural gas: Fuel for the 21st century*. John Wiley & Sons.
Staffell, I., Scamman, D., Abad, A. V., Balcombe, P., Dodds, P. E., Ekins, P., Shah, N., & Ward, K. R. (2018). The role of hydrogen and fuel cells across the global energy system. *Energy & Environmental Science, 2*, 463–491. https://doi.org/10.17605/OSF.IO/RZM4G
Stephenson, T., Valle, J. E., & Riera-Palou, X. (2011). Modeling the relative GHG emissions of conventional and shale gas production. *Environmental Science and Technology, 45*(24), 10757–10764. https://doi.org/10.1021/es2024115
Stern, J. (2017). *Challenges to the future of gas: Unburnable or unaffordable? (OIES paper NG 125)*. Oxford Institute for Energy Studies. www.oxfordenergy.org/wpcms/wp-content/uploads/2017/12/Challenges-to-the-Future-of-Gas-unburnable-or-unaffordable-NG-125.pdf
Stern, J. (2019). *Challenges to the future of LNG (OIES paper NG 152)*. Oxford Institute for Energy Studies. https://doi.org/10.26889/9781784671464
Stern, N. (2006). *The economics of climate change: The Stern review*. Cambridge University Press.
Tamburino, L., Cafaro, P., & Bravo, G. (2023). An analysis of three decades of increasing carbon emissions: The weight of the p factor. *Sustainability, 15*(4), 3245. https://doi.org/10.3390/su15043245
Tavoni, M. (2015). Natural gas and climate change policies. *Review of Environment, Energy and Economics (Re3)*. https://ssrn.com/abstract=2603182
Tusiani, M., & Shearer, G. (2016). *LNG: Fuel for a changing world: A nontechnical guide*. Pennwell Books.

UNFCCC (United Nations Framework Convention on Climate Change). (1992, May 9). *United Nations framework convention on climate change.* https://unfccc.int/files/essential_background/background_publications_htmlpdf/application/pdf/conveng.pdf

UNFCCC (United Nations Framework Convention on Climate Change). (1997, December 11). *Kyoto protocol to the United Nations framework convention on climate change.* https://unfccc.int/resource/docs/convkp/kpeng.pdf

UNFCCC (United Nations Framework Convention on Climate Change). (2015). *Paris Agreement under the United Nations framework convention on climate change.* Retrieved April 22, 2016, from https://unfccc.int/sites/default/files/english_paris_agreement.pdf

Van De Graaf, T., Sovacool, B., Ghosh, A., Kern, F., & Klare, M. (2016). *The Palgrave handbook of the international political economy of energy*. Palgrave Macmillan.

Wabnitz, C., Lam, V., Reygondeau, G., Teh, L., Al-Abdulrazzak, D., Khalfallah, M., Pauly, D., Deng Palomares, M. L., Zeller, D., & Cheung, W. W. L. (2018). Climate change impacts on marine biodiversity, fisheries and society in the Arabian Gulf. *PLoS ONE, 13*(5), e0194537. https://doi.org/10.1371/journal.pone.0194537

Wang, X., Sun, X., Ahmad, M., & Zhang, H. (2023). Does low carbon energy transition impede air pollution? Evidence from China's coal-to-gas policy. *Resources Policy, 83*, 103723. https://doi.org/10.1016/j.resourpol.2023.103723

Welsby, D., Price, J., Pye, S., & Ekins, P. (2021). Unextractable fossil fuels in a 1.5 °C world. *Nature, 597*(7875), 230–234. https://doi.org/10.1038/s41586-021-03821-8

Zhang, F., Lu, J., & Chen, L. (2023). When green recovery fails to consider coal pushback: Exploring global coal rebounds, production, and policy retrenchment post Covid-19. *Energy Research & Social Science, 101*, 103142. https://doi.org/10.1016/j.erss.2023.103142

Zhang, X., Myhrvold, N. P., Hausfather, Z., & Caldeira, K. (2016). Climate benefits of natural gas as a bridge fuel and potential delay of near-zero energy systems. *Applied Energy, 167*, 317–322. https://doi.org/10.1016/j.apenergy.2015.10.016

2 A framework for comprehensive LNG analysis

Driven by improved economics and excess supply, LNG is the fastest-growing fossil fuel. Liquid markets, standardisation of contracts, and new traders and marketers have facilitated increased trade, which in turn is increasing access to LNG supplies globally (Al-Yafei et al., 2021). Understanding LNG in a shifting global context requires a knowledge and integration of multiple contextual factors. In this chapter, I introduce a framework for the comprehensive understanding of LNG in specific contexts. A starting point to building the framework is understanding the LNG value chain.

The LNG value chain

The LNG value chain comprises five primary components. The chain starts with a series of "upstream" steps leading to the liquefaction process, including exploration and production, which usually take place in remote offshore areas. Processing of the natural gas to remove impurities and liquefaction then occurs (midstream), followed by shipping of the LNG to on-shore areas for regasification and storage. The final stage in the value chain is downstream distribution to markets (Al-Kuwari, 2023; Al-Yafei et al., 2021). Each component is discussed in turn in the following sections.

Upstream gas

Natural gas upstream production covers all the steps leading to the LNG facility. It includes components of the gas field production infrastructure and gas pipeline feeds to gas processing plants. In the gathering and boosting stage, natural gas received from multiple wells undergoes processing, including acid gas removal, dehydration, compressor operations, and the application of pneumatic devices (Roman-White et al., 2019). Processing is required to meet "sales" gas specifications that vary by region (Balcombe et al., 2015; Roman-White et al., 2019). In some cases, processed natural gas is transmitted via long-distance pipeline networks. Regular compressor stations drive the gas and help to overcome pipeline friction.

DOI: 10.4324/9781003545149-2

The composition of natural gas varies depending on the reservoirs. CH_4 features predominantly, with smaller proportions of ethane, propane, heavier alkanes, nitrogen, CO_2, and hydrogen sulfide (H_2S). In the pre-production and extraction stages, most emissions occur in (1) liquids unloading, (2) gas vented from pneumatically controlled devices, (3) fugitive emissions from flanges, (4) connectors, (5) open-ended lines and valves, and (6) natural gas combustion in operations to drive compressors and provide energy for operations (Balcombe et al., 2015; Garg et al., 2023; Roman-White et al., 2019). At the well completion stage (for unconventional wells), the hydraulic fracturing fluid returns to the surface while the gas flow increases to the initial production rate. Liquids unloading is a process to remove the flow restriction and increase production using techniques, including well blowdowns (Balcombe et al., 2015). Liquids unloading-related emissions are highly variable. Some wells do not vent during unloading, while others may be the primary source of emissions. Well age, reservoir properties, equipment, and operational strategies all can affect emissions at this stage. Most emissions come from the removal and management of impurities during processing.

The number of LNG suppliers has expanded in recent decades because of the growing demand for natural gas and the unlocking of large global reserves. In terms of technology and service providers, floating liquified natural gas (FLNG) terminals and floating storage regasification units (FSRUs) have unlocked previously uneconomic reserves and provided new opportunities (e.g. the wide deployment of hydraulic fracturing technologies enabled the production of previously uneconomic shale gas resources in North America) (Selden, 2020). However, unconventional gas requires advanced production methods and is frequently associated with environmental concerns. The US Environmental Protection Agency asserted that 66% of CH_4 emissions take place during the production and gathering phases. Unconventional gas could represent 40% of the remaining technically recoverable resource, although this comes with the caveat of high technical and economic uncertainty (McGlade et al., 2012). The United States is unique because of its combination of rich resources, high level of technical know-how, extensive domestic and readily accessible gas resource-related data, a supportive regulatory environment, including clear and transparent mineral rights ownership law, and a sophisticated and flexible financial system (Chandra, 2017; Selden, 2020). However, the potential for unconventional gas outside the United States is more uncertain: France banned the production and exploration of unconventional gas based on environmental concerns linked to the potential impacts on surface and groundwater, and the increased risk of seismic vibrations caused by underground water injection (Henriet & Schubert, 2018), and the UK faces environmental and climate-related pressures such as a focus on territorial emissions that make large-scale production of unconventional gas incompatible with long-term decarbonisation policies (McGlade et al., 2018).

Around half of global conventional natural gas resources are located in three countries – Russia, Iran, and Qatar (Moniz et al., 2011; Shayan, 2023) – with considerable unconventional gas resources identified globally. Historically, Russia and the Middle East, supported by the Organization of the Petroleum Exporting

Countries (OPEC), were the major suppliers of oil and gas, while coal suppliers were geographically more widespread. In 2015, the United States became the world's largest oil and gas producer (EIA, 2024; Energy Institute, 2024; IEA, 2018a); it is also a major energy exporter. Traditional suppliers, such as Russia and Saudi Arabia, maintain their position, competing with new suppliers from Africa, Eurasia, and Asia (IEA, 2018b).

Liquefaction

The liquefaction step includes large-scale refrigeration plants with heat exchangers for liquefying feed gas, LNG storage, and port-loading facilities. The process involves cooling the gas to approximately –160ºC and then storing it in cryogenic containers. Pre-treatment and liquefaction energy requirements are met through combusting a parasitic stream of natural gas as it leaves the pre-treatment facility, but before it enters the liquefaction facility (Roman-White et al., 2019). The average global liquefaction utilisation rate was 88.7% in 2023, compared with 74.6% in 2020, and 6 out of 21 LNG exporting markets achieved utilisation rates of more than 90% in 2020 (IGU, 2021a, 2024).

Shipping

Stored LNG is loaded onto LNG vessels, and then unloaded in storage tankers at receiving terminals containing regasification facilities. Boil-off gas is compressed and used for fuel. Historically, LNG vessel fleet growth was balanced with additional liquefaction capacity, resulting in a stable charter market, and LNG vessels that are increasingly efficient. A total of 359 vessels were under construction as of February 2024 (approximately half of the current active fleet), indicating shipowners' expectations that LNG trade will continue to grow (IGU, 2024). By the end of April 2022, there were 701 active LNG vessels, including 47 FSRUs and 10 floating storage units (IGU, 2024).

To manage air pollution, the International Maritime Organization (IMO) implemented regulations to reduce SO_X and NO_X (IGU, 2024). The industry has mostly complied by using low sulfur oil fuels, heavy foil oil with scrubbers, or switching to LNG fuel (IGU, 2021a). The IMO is considering further reductions of CO_2 emissions by 2030 and 2050 (IGU, 2021a). According to the International Gas Union (IGU, 2021a), GHG reductions of up to 23% are achievable using LNG as a maritime fuel. A growing number of markets are planning to add LNG bunkering infrastructure to meet an anticipated increase in demand for LNG. Other maritime fuels have been proposed, including ammonia and methanol, which have the advantage of being compatible with internal combustion engines. Increased electrification and challenges with scaling up CCS mean that synthetic fuels have the potential to take on a larger role.

Regasification

Regasification plants located at receiving terminals pressure LNG back into its gaseous state for distribution (IGU, 2015). Imported LNG is regasified at the receiving terminal, making it suitable for pipeline transportation and combustions. Open Rack Vaporizer technology passes the LNG through a heat exchanger to regasify it (Roman-White et al., 2019). Regasification capacity generally exceeds liquefaction capacity to meet peak seasonal demand and ensure adequate supply (IGU, 2021a). Utilisation levels depend on total market regasification capacity, infrastructure connectivity, liquidity, tradability of the wholesale gas market, competitiveness of regasification tariffs, and the attractiveness of the capacity allocation mechanism. In February 2024, average regasification utilisation levels were 41%, and global regasification capacity reached 1,029.9 million tonnes per annum (MTPA) (IGU, 2024). Regasification of the LNG via receiving terminals represents the least impactful part of the LNG chain from an emissions perspective.

Market

The final stage of the LNG value chain is the downstream connection to the natural gas transmission and distribution network that delivers gas to customers (Al-Yafei et al., 2021). Globally, there are three major gas markets: Asia, Europe, and North America. These utilise approximately 70% of gas delivered by pipeline and 30% by LNG (Ritz, 2018). LNG has restructured the natural gas industry, accounting for 37% of total gas demand in 2023 (IEA, 2024a), and it is likely to continue to play a crucial role in the industry's development and potential growth. There is also an ongoing major regional transition in the pattern of energy demand. Historically, Organisation for Economic Co-operation and Development (OECD) countries were the world's major energy consumers, representing nearly two-thirds of demand in 1990 (BP, 2019). Demand in the developing world, led by Asia, is now growing and is forecast to grow further, such that the non-OECD share is expected to represent nearly three-quarters of demand in the coming decades (ExxonMobil, 2019; IEA, 2018b).

Global LNG trade and liquefaction capacity increased, respectively, to 401.4 MTPA and 483.1 MTPA in 2023 (IGU, 2024). Despite the cancellation of nearly 160 shipments between April and November 2020 owing to COVID-19-related reduced demand, LNG was one of the few commodities that experienced growth, which is partly indicative of natural gas's resilience and flexibility in global energy systems. Indeed, global LNG trade volumes were expected to be 17% above 2019 pre-COVID-19 levels by 2024. Asia Pacific and Asia accounted for 66% of global LNG imports in 2023, with those regions' market shares increasing owing to higher demand from China and India. In 2023, the United States had the largest operational export capacity (84.5 MTPA), followed by Australia (79.6 MTPA) and Qatar (78.2 MTPA). The top three LNG exporting markets represent more than half of the global liquefaction capacity (IGU, 2021a). Russia (31.4 MTPA), Malaysia (26.8 MTPA), Indonesia (15.6 MTPA), Algeria (13.0 MTPA), and Nigeria (13.0 MTPA) were other countries contributing significantly to LNG imports in 2023 (IGU, 2024). Ouki (2023) notes that five African countries – Nigeria, Algeria, Mozambique, Egypt, and Libya – hold more than 90% of the continent's proved natural reserves, and in 2023, fewer than ten African countries were exporting gas by cross-border gas pipelines or LNG tankers.

Traditionally, LNG contracts were limited in terms of flexibility and mimicked pipelines' point-to-point trade models with dedicated shipping, take-or-pay commitments for buyers, and little to no ability for either side to divert supplies (Tusiani & Shearer, 2016). The LNG industry was characterised by capital-intensive and rigid business models that required traditional creditworthy buyers to commit to long-term contracts (Boersma & Jordaan, 2017; Bridge & Bradshaw, 2017). The initial pricing models of Alaskan and Algerian LNG were based on a fixed-price formula with escalation tied to labour and steel pricing, mirroring the US interstate pipeline tariff structures at the time (Tusiani & Shearer, 2016). For traditional LNG markets, pricing was based on oil-related indices and contracts containing long-term contractual elements such as take-or-pay commitments and dedicated volumes. Buyers were state-backed companies that could transfer the risk and costs to their end-users but who were primarily driven by volume security and playing a part in their government's energy diversification strategy (Chandra, 2017; Tusiani & Shearer, 2016).

After the OPEC oil crisis, exporters who also exported LNG insisted on oil indexation for pricing to ensure a more reliable revenue stream. Ultimately, two broad types of markets were established – liquid markets in the United States, UK, and northwest Europe with more flexible terms based on gas-to-gas competition, and oil-indexed markets, traditionally Japan and Korea, but also other regions where no competitive gas market existed. However, as global LNG supply increased, first from traditional suppliers such as Malaysia, the United Arab Emirates, and Qatar, and then from newer exporters such as Australia, production methods improved and overall project chain costs fell, making LNG a more viable option for new market entrants (Al-Yafei et al., 2021). The market evolved with multiple types of buyers in all major gas markets, moving away from the exclusive baseload LNG buyer to semi-flexible and flexible buyers, who purchase LNG based on seasonal requirements, price differentials with other fuels, or even arbitrage opportunities (Grigas, 2017; IEA, 2018b; Tusiani & Shearer, 2016). As gas and electricity markets around the world continue to liberalise, pressure will increase on buyers to improve economic terms. Supported by increased LNG volumes globally and terminals with third-party access connected to liquid markets where natural gas from a variety of sources can be processed in a bidirectional fashion, LNG markets will move towards gas-on-gas competition.

The IEA (2018b) classified LNG consumers into three categories, each with distinct demand profiles and needs – baseload LNG importers, semi-flexible seasonal LNG importers, and flexible price-sensitive LNG importers. Stern's (2017) typology for analysing potential gas markets takes into consideration each country's energy and environmental policies in relation to various gas attributes, including the replacement of domestic supply, penetration of natural gas in energy systems, affordability, and ability to switch from coal, oil, and nuclear energy. By taking Stern's methodology (2017) and applying another criterion – countries with limited or no access to pipeline gas – the potential for LNG demand expansion is clear.

This is particularly relevant for countries such as Japan, the Philippines, India, Taiwan, Indonesia, and several countries belonging to the Gulf Cooperation Council.

The state of the LNG value chain

Today, LNG industry ownership structures include players and combinations across the five key areas of exploration and production, liquefaction, shipping, regasification, and gas marketing. Historically, the full value chain contained similar participants across its length, but it is now increasingly common for liquefaction plants to be treated as separate companies, with different gas purchasing and selling arrangements (Al-Kuwari, 2023). Factors contributing to the advent of new business models include downstream integration of LNG producers and upstream integration of LNG buyers, LNG ships not being tied to specific contracts, new technological innovations – such as FSRUs and FLNG terminals, supported by more competitive financing – diversified markets, and an increase in LNG engineering, procurement, and construction companies (Tusiani & Shearer, 2016).

The LNG industry traditionally had high barriers to entry. Projects require significant investments and long-term offtake agreements from creditworthy customers. However, the US tolling model allowed a new approach whereby upstream gas producers are natural gas suppliers to LNG liquefaction companies. This decoupled the typically integrated value chain, providing opportunities for new players. Growth in the LNG sector has also attracted new entrants into shipping, liquefaction technology, and market-based tools to support LNG trade. Affordability is a key criterion determining the attractiveness of LNG in Asia, South America, and Africa (Al-Kuwari, 2023).

The number of LNG exporters and projects has increased over the past decade, and the LNG value chain is also more fragmented than ever. This fragmentation presents opportunities for new participants. With Europe and other countries moving towards net-zero, projects are competing to solidify market share. The overall economics of projects plays an important role in determining LNG exporter strategies. Established players with pre-built infrastructure benefit from economies of scale. Along with new entrants, three major exporters have clear advantages: (1) Qatar is the lowest-cost producer and has announced significant expansion plans; (2) Australia is geographically well positioned to supply Asia; and (3) before the energy crisis, Russia has indicated a strategy of market diversification to manage a move away from exported pipeline gas (Al-Kuwari, 2023).

A framework for LNG analysis

Understanding the LNG value chain and, in particular, the complexity of LNG markets is foundational for a comprehensive analysis of the role of LNG in specific contexts. Thus, the first two components of the meta-framework applied in this book derive from the LNG value chain, namely "Resources" and the "LNG Market and transactions". Resources refer to natural resources – essentially the very start of the LNG value chain – and the infrastructure required to harvest LNG

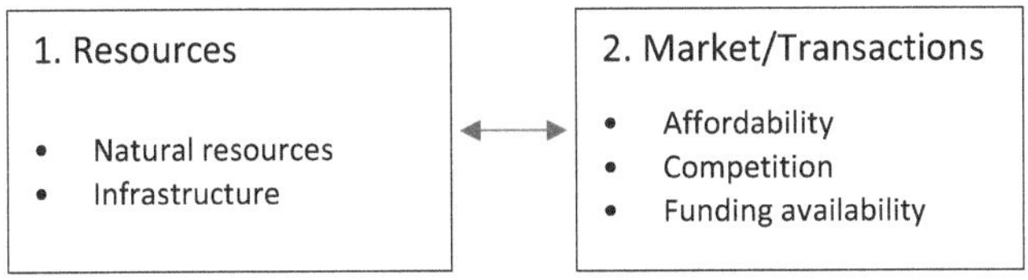

Figure 2.1 Building a framework for analysis of LNG the first two components

and bring it to market. Market refers to aspects in the downstream value chain included in the analysis, such as affordability, competition, and funding availability. Figure 2.1 shows the first part of the developing framework. The double-sided arrows between "Resources" and "Market" indicate the influence that resources have on the market, and vice versa.

A techno-economic (TE) approach analyses the economic performance of a specific technology (process, product, or service). Thus, these first two components of the framework focus mainly on TE aspects of LNG. However, socio-technical approaches to analysing transitions have gained prominence because of the recognition that energy transitions are influenced by multiple factors, and that current scientific knowledge and understanding of change is incapable of exploring the future in an interdependent system in which factors influence each other (Al-Kuwari, 2023; Fontela, 2000). There is growing recognition that stronger links are required between TE approaches (typically quantitative) and socio-technical and socio-political approaches (often qualitative) to leverage the strengths of both approaches (Fontela, 2000). Thus, I now turn my attention to the social and political environment in which LNG value chains occur. The third component of the framework relates to the regime status or dynamics of the energy transition, while the fourth component refers to power distribution or geopolitical influences.

The energy transition and the LNG industry

From an energy-transition perspective, LNG provides flexibility in planning (Al-Yafei et al., 2021). The long-evolving LNG chain, with many new players, allows for potential adaptability for the industry in a period of major transition. LNG has the potential to support the development of natural gas in different regions in the short to medium term. The industry's future depends on this ability to provide flexible solutions, while realising medium-term trends of cost reduction and technological innovation. Longer term, the major challenge will be the impact of decarbonisation policies. Steuer (2019, p. 255) noted that the "real challenge of the future of gas is more directly linked with energy policy formulation and the speed of global decarbonisation rendering natural gas unburnable before it becomes unaffordable". Decarbonisation technology, in particular CCS and hydrogen, will determine the long-term viability of natural gas, including LNG.

The LNG industry is credited with changing the role of natural gas towards a more globalised commodity (Colombo et al., 2016). Initially predicted to be world's largest LNG importer, today the United States is the fastest export growth region (Corbeau & Ledesma, 2016; Energy Institute, 2024; IEA, 2024b; IGU, 2024), driven largely by the shale gas revolution. The US shale gas revolution accelerated the transformation of the LNG industry by decoupling the upstream and midstream parts of the LNG chain, where liquefaction plants are positioned as customers of natural gas markets (Selden, 2020; Tusiani & Shearer, 2016). The impact of the United States as an LNG exporter cannot be overstated. US LNG introduced new elements of flexibility and diversity into the LNG market. Propelled by abundant domestic natural gas supplies, and modern and efficient production methods, US LNG introduced a wide range of LNG players into the market and cemented itself as a potential competitively priced source of LNG in the long term. Natural gas exporters, LNG export facilities, shipping firms, traders, and developers all are participating in the fragmented US LNG ecosystem. Thus, in addition to diversifying global supply, US LNG has developed more flexible commercial approaches with new dynamics that bring shorter contract terms to the market and pricing mechanisms that are more responsive to global price signals. LNG buyers now have access to a significant amount of flexible LNG that can be adapted to requirements, including seasonal markets, short-term energy shortages, and power-to-gas applications.

In Japan, the Fukushima disaster demonstrated LNG's flexibility in meeting high-volume, short-term demand. However, it also exposed Japan to pricing volatility, which had a significant economic impact. Post-Fukushima, both Japan and South Korea have been hesitant to continue their heavy dependence on natural gas by accelerating energy efficiency and renewable plans while expanding coal-generated capability despite ambitious climate goals (Boersma & Jordaan, 2017). LNG has evolved from a point-to-point model to a diverse, complex, and emergent global market, growing in scale and importance (Bridge & Bradshaw, 2017).

Natural gas is associated with several uncertainties, such as the competitiveness of other fuels and global decarbonisation policies. However, LNG has characteristics such as operational and commercial flexibility, diversification, and technological advances that may set it on a different path. LNG demand depends on the future role of gas in the energy mix, but that role will be partly determined by the success of LNG. LNG offers a solution that may provide additional benefits over pipeline gas. The LNG industry is growing, primarily because of its flexibility and geopolitical considerations. LNG exporters must consider whether this trend is sustainable or if countries are using it as a flexible transition fuel until lower-carbon options are available. Previously fixed vertical supply chains are now decentralised and disaggregated with buyers, traders, and global portfolio owners having a role in the market. Supported by readily available financing, freedom from long-term contracts, different ownership structures of vessels and terminals, and the growth of liquid markets such as the Henry Hub in the United States and the UK's National Balancing Point, the risk and reward profile has fundamentally changed the attractiveness of LNG versus pipeline gas. However, emissions remain an important consideration for both LNG and pipeline gas.

Emissions

Shipping distance and vessel technology play a major role in the LNG segment, and CH_4 and CO_2 are the major GHG emissions along the chain. Upstream activities are responsible for a larger share of emissions owing to the greater role of CH_4 emissions in that part of the chain. Significant variance exists in total CO_2e intensity across life-cycle comparisons in the segments of the LNG value chain. The main variances arise in the extraction phase (including production, pipeline transmission, and gathering and boosting) owing to the technologies used, resource type, and state and age of the infrastructure. There is generally more consistent data in LNG-specific processes because of less fugitive CH_4 and greater visibility of emissions data and processes.

CH_4 is particularly important given its impact on global warming and increased scrutiny from environmental groups and policymakers, specifically in the production and transportation segments. The global warming potential (GWP) time applied to CH_4 (20 or 100 years) can play a significant role in driving variances. In all scenarios, power generation accounts for most emissions, and, in most cases, the LNG or natural gas supplier does not manage this segment. Published estimates of CH_4 emissions from the natural gas value chain vary widely because of (1) the magnitude and range of emissions across supply chains, (2) different methods, data, and assumptions used, and (3) the timescale of GWP of CH_4 (Balcombe et al., 2015; Roman-White et al., 2019; Stern, 2020). Super emitters – a small number of high-emitting facilities with ineffective processing equipment and poor operational strategies – notably skew the reporting of CH_4 emission profiles (Balcombe et al., 2015; IEA, 2024b; Schuit et al., 2023; Stern, 2020).

Most CH_4 emissions take place outside LNG-specific processes. They occur mainly in the production, processing, and transportation of natural gas. In the production stage, fugitive and vented CH_4 from pneumatic devices, leaks, and liquids unloading are major contributors. During these processes, natural gas is dehydrated with glycol as a solvent. When the liquid is flushed to regenerate the solvent, CH_4 and other hydrocarbons dissolved in it are discharged. Hydrocarbons are also released from the vent. They are burned by flaring before being discharged or are brought back into the fuel line to decrease the number of hydrocarbons released into the air (Daudu et al., 2024; Tamura et al., 2001). In the transport phase, CH_4 escapes from reciprocating compressor drivers and from compressor venting (Roman-White et al., 2021).

GHG levels from LNG supply reached 1.35 GT CO_2e in 2020, representing approximately 17% of all natural gas emissions (IEA, 2021a). Accurately evaluating GHG emissions of LNG is, however, complex. Calculations depend on several variables and assumptions across the entire chain. GHG emission factors, emission rates, process technology choices, carbon intensity of the feedstock gas, and operational conditions are all factors that affect emission profiles (Balcombe et al., 2015, 2017; Daudu et al., 2024; Gao et al., 2023; IGU, 2015; Pace Global, 2015; Roman-White et al., 2019, 2021). Furthermore, the prominence of the United States as a major gas producer introduced several complexities into GHG emission analysis, along with increased scrutiny of CH_4 leaks from natural gas production systems as a result of the US shale boom (Selden, 2020). Determining GHG emissions – particularly CH_4 emissions – for LNG exported from the United States is difficult because of the need to identify the field/s where the gas originates from and the specific pipeline route taken to the LNG terminal (Stern, 2020; Wettengel, 2020).

Compared with oil and coal value chains, LNG's indirect emissions are larger because of (1) the comparatively large amount of energy required to liquefy natural gas, (2) emissions released during natural gas extraction, and (3) LNG processing and transport (Daudu et al., 2024; Hondo, 2005; Simpa et al., 2024). Moreover, the net emissions impact of increased LNG in a country's electricity system should be assessed while taking into consideration the importer country's power mix, marginal sources of new domestic power, market dynamics, and available end-uses (Boersma & Jordaan, 2017). Three components of the supply chain stages of the overall natural gas value chain – upstream, LNG chain, and pipeline gas – allow a better understanding of the emissions impact at each stage. In the pre-treatment process before liquefaction, CO_2 is removed from natural gas using solvents such as amines. In this solvent regeneration process, CO_2 and CH_4 are released into the air as off-gas (Tamura et al., 2001). CO_2 is also discharged when natural gas is consumed as fuel by gas turbines that power liquefaction plants. Fuel consumption depends on the productive capacity and efficiency of liquefaction plants (Tamura et al., 2001). Many of Asia Pacific's LNG facilities are in remote areas, far from the power grid. As a result, feed gas is used to generate electricity to run the plant and fuel the liquefaction process. Typically, 8–12% of feed gas is consumed at the plant to run these processes. Older, more inefficient plants, and nascent FLNG vessels operate with far higher losses (Wood Mackenzie, 2020).

In the shipping process, fuel consumption and boil-off gas are responsible for most emissions. Thus, shipowners face significant challenges to manage CO_2 emissions. From 2023, ships above 400 gross tonnage must comply with new standards of energy efficiency required by the Energy Efficiency Existing Ship Index (IGU, 2021b), mainly by limiting ship propulsion power, the primary fuel consumer.

The main types of emissions in the LNG value chain are combustion emissions, fugitive emissions, and vented and flared emissions (sometimes not captured by a system owing to equipment leaks and other processes). These types should not be confused with sources of emissions such as liquefaction, regasification, storage, and transport (Balcombe et al., 2015; Roman-White et al., 2019). Parts of the chain require heat or electricity to operate, and demand is typically met by natural gas. Mostly, this results in CO_2 emissions, although CH_4 can also be emitted when there is incomplete combustion (Balcombe et al., 2015; Simpa et al., 2024; Tamura et al., 2001). Both CH_4 and CO_2 must be vented at various points along the LNG value chain, mainly in the production stages. When feasible, capturing or flaring of vented gas is the norm. Purge gas – used in emergencies or in start-up – is burned and discharged from the flare stack in the form of CO_2. In normal operations, some gas fields release no CO_2 from flare combustion (Balcombe et al., 2015; Tamura et al.,

2001). Fugitive emissions (unintentional gas leaks) typically occur during the extraction and pipeline transport phases.

LNG versus pipeline gas

Several variables must be considered when comparing the environmental impact of LNG versus pipeline gas. In a like-for-like scenario, where both the source and consumption pattern of gas are identical, the difference will be in emissions related to shipping, regasification, pipeline transportation, and LNG liquefaction. On average, LNG results in higher emissions given the additional steps of liquefaction, tanker transport, and regasification (Anderson & Bows, 2011; Stephenson et al., 2012). The LNG chain is also generally more environmentally damaging than the pipeline gas chain, mainly because of the additional power needed to liquefy and transport LNG (Anderson & Broderick, 2017; Boersma & Jordaan, 2017; De Jong, 2023; Gilbert & Sovacool, 2018).

While pipeline gas and LNG deliver the same product, their market characteristics differ because of their associated physical and commercial factors. Traditionally, the decision to identify the appropriate way to commercialise and export gas was based on three factors – distance between gas source and market, exportable volumes, and feed gas cost (Chandra, 2017; Colombo et al., 2016). The industry considered distances of greater than 2,000–2,500 km between gas field and buyer as indicative of consideration of LNG over pipeline gas. This also depended on the geography, geopolitical status of transit countries, and overall commercial terms (Tusiani & Shearer, 2016). In terms of volume, a 1-million-tonne-per-annum LNG project requires 1 trillion cubic feet (TCF) of exportable gas (Chandra, 2017). The cost of feed gas depends on many factors, including gas quality, the degree of treatment and processing required, purchase terms with the resource owner, and contractual-term commitments to ensure long-term project economics. Table 2.1 summarises key differences and strategic considerations for pipeline gas and LNG.

CH_4 emissions and CO_2 content affect overall emissions when comparing LNG with alternatives. While difficult to generalise, comparison factors include the feed gas source, types of end-use in importing countries, CH_4 leakage rate, and each importing country's energy policy (Boersma & Jordaan, 2017; Gilbert & Sovacool, 2018). In a study analysing the climate impact of a range of Australian energy sources exported to China, Hardisty et al. (2012) demonstrated the benefits of US and other exported LNG for power generation over domestic coal use in Asia and Europe. In contrast, Gilbert and Sovacool (2017) concluded that US LNG exports were likely to increase global GHG emissions through increased consumption, higher emissions in the United States, and CH_4 leakage.

With impurities removed prior to liquefaction, the delivered LNG is usually composed of purer CH_4 with a higher overall energy content (Kumar et al., 2011). This can be a problem for buyers accustomed to receiving leaner gas (Kavalov et al., 2009). The CO_2 content of delivered LNG is also generally lower than that of pipeline gas. Calculations by Wood Mackenzie (2017) show that approximately 10% less CO_2 is present in the natural gas combusted downstream from LNG

Table 2.1 Pipeline gas versus LNG strategic considerations (Al-Kuwari, 2023)

	Fuel source	*Full chain emissions*	*CO_2 content (delivered gas)*	*Short-term flexibility (on the day)*	*Medium-term flexibility (seasonal market)*	*Synergies with renewable energy*	*Carbon lock-in*	*Decarbonisation*
Pipeline gas	Natural gas	Source of natural gas; length of pipeline emissions (fugitive and flaring)	No CO_2 extraction as part of the process	Access to pipeline gas storage (highly flexible)	Commercial terms; physical capacity of pipeline	In day flexibility (highly compatible); seasonality leads to variation in use	Pipeline investment (long term/ significant); natural gas processing (long term)	CCS at source (hydrogen); carbon capture and processing (with power)
LNG	Natural gas	Pipeline emissions to upstream LNG; efficiency of LNG process; shipping and regasification emissions (fugitive and flaring)	CO_2 extracted as part of LNG process (up to 10% on average)	Access to stored LNG (limited)	Commercial terms; LNG terminal capacity (highly flexible)	In day flexibility limited by LNG storage (low compatibility); seasonality leads to planning for periods of low/high demand (i.e. heating highly compatible)	LNG terminal (long term/ significant) FSRU or FLNG (short term/ minimum)	CCS at source and export (hydrogen or ammonia) or hydrogen CCS at LNG terminal (hydrogen). Potential to leverage CO_2 extraction as part of LNG process

sources than in pipeline gas. The majority of LNG projects – with some exceptions, such as Australia's Gorgon LNG (Clark & Herzog, 2014) – vent CO_2 rather than capturing it.

In the long term, if LNG exporters can capture the vented CO_2 and any CO_2 produced from liquefaction and shipping, overall emissions comparable to most modern pipeline gas projects can be minimised, and in some cases even provide an environmental advantage. CO_2 emissions can also be reduced in liquefaction facilities using carbon capture technologies. Coulson et al. (2010) concluded that while 0.02 tonne CO_2/tonne LNG can be reduced by CCS, heat recovery and integration optimisation options provide more cost-effective solutions. LNG facilities with feed gas containing high CO_2 content will emit more CO_2 unless the carbon is captured and sequestered. The Snøhvit LNG plant in Norway captures nearly 700,000 tonnes of CO_2 a year, which is transported 145 km offshore to a depth 2,500 m below the seabed, and the Gorgon and Browse plants in Australia seek to follow a similar model (Coulson et al., 2010). Ninety per cent of CO_2 emissions related to LNG liquefaction are the result of fuel combustion, but significant CO_2 reductions can be achieved by addressing combustion efficiency by end-users, depending on plant configuration, feed composition, and liquefaction technology.

Tackling fugitive CH_4 emissions and the extraction process of the source gas shifts the assessment of the overall environmental impact of LNG versus pipeline-delivered gas in favour of LNG (Lin & Brooks, 2021). Europe and China have the greatest gas-on-gas competition between LNG and gas pipeline imports (Wood, 2024). Both regions play a balancing market role owing to their large pipeline and LNG import infrastructure, which can absorb large amounts of supply (Corbeau & Ledesma, 2016; BP, 2019). Security and flexibility of supply are attractive points for LNG, and the industry benefits from these as it increasingly seeks to commoditise and depoliticise the gas trade (Daudu et al., 2024; Grigas, 2017; IEA, 2018b). The growth of LNG infrastructure, from the upstream process to shipping and regasification terminals, provides the fundamentals for a truly global market similar to the emergence of global oil markets.

Natural gas demand as part of the global energy mix

The electricity sector is forecast to experience dramatic change as the fastest-growing energy carrier in all scenarios. The push towards electrification is mainly attributable to global climate mitigation policies and the drive to decarbonise energy systems. Continued cost reductions in solar and wind energy, supported by advances in storage technology, also make renewables easier to integrate into power grids (BP, 2019, 2022; IEA, 2018b, 2021b, 2021c, 2023a, 2023b, 2024b). Most electricity growth is occurring in developing economies, with electric transportation and increased demand for cooling being the main drivers. Energy demand for buildings is also growing through improved economic conditions in developing economies. Despite significant growth, in most scenarios renewable energy is unable to meet all additional power demand, and there is a general increase in natural gas consumption in the power sector in the short and medium term (BP, 2019, 2022; IEA,

2018b, 2021b, 2021c). The increased demand is forecast to be met by natural gas for the next few decades, although electricity is likely to penetrate the sector, and hydrogen will also play a role (Bloomberg, 2020; BP, 2019, 2022; IEA, 2018b, 2024b; IRENA, 2018; Royal Dutch Shell, 2018).

From a demand perspective, China (see Chapter 5) and India will play a significant role in determining the growth of gas. Both countries face challenges in terms of balancing local pressure to combat air pollution and the global fight against climate change with energy demand growth. As is the case for China, air quality in India is a key concern. However, the Indian market is more price-sensitive and faces challenges in building the infrastructure for further gas developments (Chandra, 2017; Corbeau & Ledesma, 2016; IEA, 2018b). Gas demand in the industrial sector has a clearer path in some scenarios (ETC, 2018) but must challenge renewables and domestic coal for market share in the world's fast-growing electricity sector.

Multi-level perspective analysis

Energy and sustainability transitions are multidimensional by nature and affected by many highly uncertain and complex variables (McDowall, 2014; Turnheim et al., 2015). The multi-level perspective (MLP) is a qualitative framework that combines concepts from evolutionary economics, the sociology of innovation, and neo-institutional theory (Geels, 2004, 2014; Geels et al., 2016b) to expand on transition ideas found in policy, interpretivism, and other fields. The framework can be applied to any socio-technical transition – for instance, transport, housing, and communication – not just energy (Geels, 2005). It does not attempt to incorporate all social theories but rather focuses on how new innovations, also known as "emerging niches", compete against established socio-technical regimes that benefit from path dependence and lock-in (Geels et al., 2016b). The MLP framework stipulates that transitions occur if pressure is exerted on the established regime to open new niche opportunities. Such policy-protected spaces allow for new ideas and innovations to exist, break through, and replace existing technologies and structures (Andrews-Speed, 2016; Van de Graaf et al., 2016). A new regime can be established following a period of major restructuring that may be characterised by changes in policy, user behaviour and practices, and infrastructure (Geels, 2004).

The MLP framework allows reasons for tensions and misalignment to be highlighted. At the highest level, regimes exist within a landscape of geographical, demographic, political, cultural, and behavioural norms. Pressures can emerge either globally or locally and may be triggered by changes in society, the economy, or the emergence of new beliefs or ideas (Andrews-Speed, 2016). For instance, increased awareness of the negative impacts of climate change is driving changes in culture and human behaviour and triggering public policy and innovation changes in transportation, air quality, and recycling (Geels, 2004). In the MLP framework, socio-technical transitions occur when innovations at the niche level gain momentum, exogenous events or developments pressure the regime, and tensions within the regime create opportunities for innovations to flourish (Geels et al., 2016b).

A regime transition is a profound change that spans societies and organisations. Such a transition is successful if the pressures from the different levels are mutually reinforcing (Andrews-Speed, 2016; Meadowcroft, 2009).

Transition studies are well suited to the analysis of path dependence and carbon lock-in phenomena, especially in the context of how low-carbon technologies deal with established socio-technical systems as part of the ongoing energy transition (Eitan & Hekkert, 2023; Fuchs et al., 2012; Geels et al., 2016a). In socio technical systems, rules and regimes offer stability by providing a structure of interdependence between organisations in which actors are embedded. This leads to increased investment, economies of scale, and linkages that make it difficult to change socio-technical systems (Geels, 2004).

Smith et al. (2005) developed a method to assess the adaptive capacity of a regime, including the creation of new knowledge through public research and development (R&D) programmes, the supply of resources through grant programmes, and the formation of markets through government procurement and fiscal policies (Jacobsson & Johnson, 2000; Smith et al., 2005). The adaptive capacity referred to in this model has emerged in the climate change literature (IPCC, 2001) and builds on notions of vulnerability and resilience in ecology. Regimes use their capacity and resources to respond to selection pressures (Smith et al., 2005). Andrews-Speed (2016, p. 218) defined adaptive capacity as the "ability to respond to challenges posed by changes in its environment" and its capability can be determined by its "material resources, infrastructure, information technology, human capital, wealth, financial resources and institutions and governance". North et al. (2009) referred to adaptive efficiency as a society's ability to learn and innovate through experimentation, and Smith et al. (2005) determined adaptive capacity to be a key concept used in their framework to assess a regime in an energy transition. Four contexts are described based on a division of two categories – highly coordinated contexts of endogenous renewal and purposive transition, and low coordinated context or reorientation of trajectories and emergent transformation (Smith et al., 2005).

Thus, for the country analyses that comprise the main body of this work, an MLP analysis (Geels, 2002) will establish the landscape and niche-level dynamics that can alter the course of the existing regime, with a particular focus on the current and potential future innovation and R&D ecosystem. Subsequently, the transition context will be evaluated using the methodology of Smith et al. (2005), which maps out a country's resources and its ability to respond to the energy transition; this will help to assess potential pathways for the country's energy future and to analyse what might affect the current natural-gas-dependent system in the context of decarbonisation and other policy goals. Each country will be located within one of Smith et al.'s (2005) quadrants, and adaptive capacity and selection process will be tabulated for each one.

Power and geopolitical influence

Climate change means that governments will have to play national and international roles to intervene in markets to manage the transition away from fossil fuels

(Al-Kuwari, 2023; Mazzucato, 2018). In addition, innovation chains need to be supported and managed because consumers are likely to be unwilling to pay for the indirect benefits of renewable energy (Fouquet, 2010; Fouquet & Pearson, 2012). Government support for energy transition can go beyond technological support, as demonstrated in the Dutch model of creating capabilities and platforms for coordination (Kemp, 2010). Grübler (2012) argued that the importance of energy end-use and its direct impact on consumers in society will be a significant challenge to the transition. However, as the comparative advantages of new services emerge, perceptions may change. Advantages may be traditional ones, such as price and reliability, but they could also take the form of modern services linked to the circular economy and overall convenience. Lifestyle changes can be a factor in spurring economic and innovation growth, driven by a "new smart, green way of living" (Perez & Leach, 2018, p. 1).

Several countries have taken a proactive approach to the energy transition, albeit with mixed results. The Netherlands developed policies to support industrial growth by creating an entity within the Ministry of Economic Affairs in 2002 (Kemp, 2010). Although it has successfully integrated several components into a development plan to deal with climate change, the approach yielded limited impacts on energy policy (Meadowcroft, 2009). Germany offers another example of a policy-driven national plan, with its bold push to a low-carbon energy system called *Energiewende*. The situation was made more complex by interest groups operating within Germany, whose requirements included a radical policy push towards full-scale renewable energy electrification and the maintenance of domestic coal production with carbon capture technology. The different interests within Germany provide an example of the challenges posed by the energy transition – despite a clear national policy, pathways to curb emissions while meeting additional demand are difficult to develop and implement (Fuchs et al., 2012; Schiffer & Trüby, 2018). China's government took a more aggressive approach towards a low-carbon transition, recognising that its future depends on its ability to improve air quality, reduce overall pollution, effectively manage natural resources, and be a major manufacturer of low-carbon technology (Mazzucato, 2018). The impact of COVID-19 – reflected in energy markets by the recent collapse in the price of US oil futures – indicates how quickly our world can change. Furthermore, war and regional conflict have global and regional implications for LNG and the energy transition.

Geopolitical concerns related to security of supply have been driven primarily by events in the pipeline gas industry, based on exposure in transit countries and the inherent leverage gas producers have over consumers in a gas pipeline arrangement. The Russia/Ukraine war led to an energy crisis with an enduring impact in Europe in terms of dependence on Russian pipeline gas (Dahan & Altheeb, 2023; De Jong, 2023; IEA, 2023b, 2024a; IGU, 2024; Selden, 2020). The conflict led to gas disruptions and increased the politicisation of the gas trade (Colombo et al., 2016; Dahan & Altheeb, 2023; Selden, 2020). Ritz (2018, p. 2) noted that pipelines "have a high degree of asset specificity" and that, once built, they are physically bound to a fixed route. Geopolitical concerns have pushed gas-importing countries to rethink their strategies and consider investments in LNG importing facilities. Ouki (2023) describes a "temporary wave of political enthusiasm for African gas" (p. 1), noting that the European interest in LNG from Africa has been welcomed by many African countries and given African policymakers the opportunity to draw attention to the limited flow of international hydrocarbon investment in Africa.

LNG was historically seen as a last choice option because of its high cost. However, security advantages often override economic considerations (Tusiani & Shearer, 2016) – particularly for traditional well-established Asian buyers who value the lack of disruption and the ability to receive large volumes when needed, as demonstrated in Japan in the aftermath of the 2011 Fukushima nuclear accident (Corbeau & Ledesma, 2016; and see Chapter 4). Japanese buyers needed to replace a significant portion of the country's electricity generation in a limited period of time, affecting overall domestic energy prices (Len & Nian, 2014). Countries are willing to pay a premium for LNG to mitigate strategic risks and create more supply options, including overinvesting in import facilities to maintain market optionality and create a buffer to manage demand fluctuations. This is accentuated by the construction of regasification terminals despite international utilisation rates of 33% between 2010 and 2015 (Colombo et al., 2016). Although short-term high volatility may have a negative impact on natural gas's role for importers, the flexibility demonstrated by LNG exporters helps manage sudden demand changes.

Energy imports are viewed as insecure compared with domestic energy sources (Stern, 2017). In terms of gas imports, therefore, countries have different considerations when determining their purchasing strategy. Buyers with no access to pipeline gas who depend exclusively on LNG imports must determine the appropriate mix of LNG in their energy supply and maximise security by diversifying supply sources or focusing on exporters deemed more geopolitically and technically secure. This must be balanced by economic considerations and the overall competitiveness of the imported LNG. For gas buyers with the option between pipeline and LNG, the decision is between market and physical security, while considering the political risks (Bridge & Bradshaw, 2017; Stern, 2017).

The T-E-T-I framework

The T-E-T-I framework ensures that all key aspects are considered when conducting an in-depth analysis of LNG in specific context. The literature review on which the framework is based and a more detailed outline of the development of the framework can be found in Al-Kuwari (2023). Given the complexity of the analysis required for energy-transition dynamics and power, additional theories and models are incorporated into the framework.

The framework combines elements of three approaches used to study transitions – techno-economic (TE), transition (T), and institutional (I) approaches, giving rise to the acronym for the framework – the T-E-T-I framework. TE considerations relate to how energy decisions are made, often centring around analyses of various integrated assessment models and energy-specific systems. Models have become better at accounting for certain socio-technical aspects – such as technology learning and spatial diffusion issues – but matters of governance and power structures are often not appropriately integrated. Al-Yafei et al. (2021) undertook a systematic review of the sustainability development of the LNG industry, describing 168 studies conducted between 2010 and 2020 that met the review criteria. The authors point out the lack of studies that holistically consider the entire LNG value chain along with environmental, economic, and social impacts. Thus, the T-E-T-I framework was developed to address the need for a more comprehensive framework and presents an opportunity to analyse the institutional, political, and resource characteristics that affect decision-making over natural gas and LNG demand in various energy markets. The framework provides a tool to analyse and compare the four areas of focus and is applied in the following chapters that focus on LNG in specific contexts. Analyses using these criteria have the potential to plug the existing gaps in our understanding of energy decisions, particularly those related to LNG. In the following chapters, the T-E-T-I model is applied to three countries to advance our understanding of LNG in those settings.

References

Al-Kuwari, O. (2023). *The future of liquified natural gas (LNG) in the energy transition: Options and implications for the LNG industry in a decarbonising world* [Unpublished doctoral dissertation]. University College London. https://discovery.ucl.ac.uk/id/eprint/10173650/

Al-Yafei, H., Aseel, S., Kucukvar, M., Onat, N. C., Al-Sulaiti, A., & Al-Hajri, A. (2021). A systematic review for sustainability of global liquified natural gas industry: A 10-year update. *Energy Strategy Reviews*, *38*, 100768. https://doi.org/10.1016/j.esr.2021.100768

Anderson, K., & Bows, A. (2011). Beyond "dangerous" climate change: Emission scenarios for a new world. *Philosophical Transactions of the Royal Society A: Mathematical, Physical and Engineering Sciences*, *369*(1934), 20–44. https://doi.org/10.1098/rsta.2010.0290

Anderson, K., & Broderick, J. (2017). *Natural gas and climate change*. University of Manchester. https://research.manchester.ac.uk/en/publications/natural-gas-and-climate-change

Andrews-Speed, P. (2016). Applying institutional theory to the low-carbon energy transition. *Energy Research and Social Science*, *13*, 216–225. https://doi.org/10.1016/j.erss.2015.12.011

Balcombe, P., Anderson, K., Speirs, J., Brandon, N., & Hawkes, A. (2015). *Methane and CO2 emissions from the natural gas supply chain: An evidence assessment*. Imperial College London. https://vibe.cira.colostate.edu/ogec/docs/reports/SGI_White_Paper_methane-and-CO2-emissions_WEB-FINAL.pdf

Balcombe, P., Anderson, K., Speirs, J., Brandon, N., & Hawkes, A. (2017). The natural gas supply chain: The importance of methane and carbon dioxide emissions. *ACS Sustainable Chemistry and Engineering*, *5*(1), 3–20. https://doi.org/10.1021/acssuschemeng.6b00144

Bloomberg. (2020). *Hydrogen economy outlook – key messages*. BloombergNEF. https://data.bloomberglp.com/professional/sites/24/BNEF-Hydrogen-Economy-Outlook-Key-Messages-30-Mar-2020.pdf

Boersma, T., & Jordaan, S. M. (2017). Whatever happened to the golden age of natural gas? *Energy Transitions*, *1*(2), 1–8. https://doi.org/10.1007/s41825-017-0005-4

BP (British Petroleum). (2019). *BP energy outlook 2019*. www.bp.com/content/dam/bp/business-sites/en/global/corporate/pdfs/energy-economics/energy-outlook/bp-energy-outlook-2019.pdf

BP (British Petroleum). (2022). *BP energy outlook 2022*. www.bp.com/content/dam/bp/business-sites/en/global/corporate/pdfs/energy-economics/energy-outlook/bp-energy-outlook-2022.pdf

Bridge, G., & Bradshaw, M. (2017). Making a global gas market: Territoriality and production networks in liquefied natural gas. *Economic Geography*, *93*(3), 215–240. https://doi.org/10.1080/00130095.2017.1283212

Chandra, V. (2017). *Fundamentals of natural gas: An international perspective*. PennWell Books.

Clark, V. R., & Herzog, H. J. (2014). Can stranded fossil fuel reserves drive CCS deployment? *Energy Procedia*, *63*, 7261–7271. https://doi.org/10.1016/j.egypro.2014.11.762

Colombo, S., El Harrak, M., & Sartori, N. (Eds.). (2016). The future of natural gas: Markets and geopolitics. *European Energy Review*. https://doi.org/10.1016/0040-1625(87)90047-3

Corbeau, A., & Ledesma, D. (Eds.). (2016). *LNG markets in transition: The great reconfiguration*. Oxford University Press for the Oxford Institute for Energy Studies and the King Abdullah Petroleum Studies and Research Center.

Coulson, M., Ferguson, S., Bullen, T., & Wheeler, F. (2010). Carbon capture options for LNG liquefaction. *Proceedings of the Sixteenth International Conference and Exhibition on Liquefied Natural Gas*, 18–21. https://doi.org/10.1016/j.enbuild.2018.02.040

Dahan, A., & Altheeb, R. (2023). Can the Arab's natural gas secure the Europeans' gas requirements? The case of liquified natural gas (LNG). *Uncertain Supply Chain Management*, *11*(4), 1677–1684. https://doi.org/10.5267/j.uscm.2023.7.001

Daudu, C. D., Adefemi, A., Adekoya, O. O., Okoli, C. E., Ayorinde, O. B., & Darajimba, A. I. (2024). LNG and climate change: Evaluating its carbon footprint in comparison to other fossil fuels. *Engineering Science & Technology Journal*, *5*(2), 412–426. https://doi.org/10.51594/estj/v5i2.803

De Jong, M. (2023, January). *LNG: Saviour or a new problem in the making? GIES occasional paper, the global energy crisis*. Ghent Institute for International and European Studies. www.ugent.be/ps/politiekewetenschappen/gies/en/research/publications/gies_papers/2023-global-energy-crisis/pdf-files/lng-saviour-or-a-new-problem-in-the-making-moniek-de-jong.pdf

EIA (Energy Information Administration). (2024, March 11). *United States produces more crude oil than any country, ever*. www.eia.gov/todayinenergy/detail.php?id=61545

Eitan, A., & Hekkert, M. P. (2023). Locked in transition? Towards a conceptualization of path-dependence lock-ins in the renewable energy landscape. *Energy Research & Social Science*, *106*, 103316. https://doi.org/10.1016/j.erss.2023.103316

Energy Institute. (2024). *Statistical review of world energy* (73rd ed.). www.energyinst.org/__data/assets/pdf_file/0006/1542714/EI_Stats_Review_2024.pdf

ETC (Energy Transitions Commission). (2018). *Mission possible: Reaching net-zero carbon emissions from harder-to-abate sectors by mid-century*. www.energy-transitions.org/publications/mission-possible/

ExxonMobil. (2019). *Outlook for energy: A perspective to 2040*. www.ief.org/_resources/files/events/ief-lecture-exxonmobils-2019-outlook-for-energy-a-perspective-to-2040/exxonmobil-2019-outlook-for-energy-a-perspective-to-2040.pdf

Fontela, E. (2000). Bridging the gap between scenarios and models. *Foresight*, *1*(2), 5–9. http://dx.doi.org/10.1108/14636680010802447

Fouquet, R. (2010). The slow search for solutions: Lessons from historical energy transitions by sector and service. *Energy Policy*, *38*(11), 6586–6596. https://doi.org/10.1016/j.enpol.2010.06.029

Fouquet, R., & Pearson, P. (2012). Past and prospective energy transitions: Insights from history. *Energy Policy*, *50*, 1–7. https://doi.org/10.1016/j.enpol.2012.08.014

Fuchs, G., Hinderer, N., Kungl, G., & Neukich, M. (2012). Adaptive capacities, path creation and variants of sectoral change: The case of the transformation of the German energy supply system Stuttgarter. *Institute for Social Sciences Organizational Sociology and Innovation Studies*, *2*, 1–49. https://d-nb.info/1191431711/34

Gao, S., Zhang, G., Guan, C., Mao, H., Zhang, B., & Liu, H. (2023). The expansion of global LNG trade and its implications for CH_4 emissions mitigation. *Environmental Research Letters*, *19*, 014022. https://doi.org/10.1088/1748-9326/ad0eff

Garg, S., Boz, D. E., Gilbert, B., & Crompton, J. (2023). A critical review of natural gas emissions certification in the United States. *Environmental Research Letters*, *18*(2), 023002. https://doi.org/10.1088/1748-9326/acb4af

Geels, F. W. (2002). Technological transitions as evolutionary reconfiguration processes: A multi-level perspective and a case-study. *Research Policy*, *31*(8–9), 1257–1274. https://doi.org/10.1016/S0048-7333(02)00062-8

Geels, F. W. (2004). From sectoral systems of innovation to socio-technical systems: Insights about dynamics and change from sociology and institutional theory. *Research Policy*, *33*(6–7), 897–920. https://doi.org/10.1016/j.respol.2004.01.015

Geels, F. W. (2005). The dynamics of transitions in socio-technical systems: A multi-level analysis of the transition pathways from horse-drawn carriages to automobiles. *Technology Analysis & Strategic Management*, *17*(4), 445–476. https://doi.org/10.1080/09537320500357319

Geels, F. W. (2014). Regime resistance against low-carbon transitions: Introducing politics and power into the multi-level perspective. *Theory, Culture & Society*, *31*(5), 21–40. https://doi.org/10.1177/0263276414531627

Geels, F. W., Berkhout, F., & Van Vuuren, D. P. (2016a). Bridging analytical approaches for low-carbon transitions. *Nature Climate Change*, *6*(6), 576–583. https://doi.org/10.1038/nclimate2980

Geels, F. W., Kern, F., Fuchs, G., Hinderer, N., Kungl, G., Mylan, J., Neukirch, M., & Wassermann, S. (2016b). The enactment of socio-technical transition pathways: A reformulated typology and a comparative multi-level analysis of the German and UK low-carbon electricity transitions (1990–2014). *Research Policy*, *45*(4), 896–913. https://doi.org/10.1016/j.respol.2016.01.015

Gilbert, A. Q., & Sovacool, B. K. (2017). US liquefied natural gas (LNG) exports: Boom or bust for the global climate? *Energy*, *141*, 1671–1680. https://doi.org/10.1016/j.energy.2017.11.098

Gilbert, A. Q., & Sovacool, B. K. (2018). Carbon pathways in the global gas market: An attributional lifecycle assessment of the climate impacts of liquefied natural gas exports from the United States to Asia. *Energy Policy*, *120*, 635–643. https://doi.org/10.1016/j.enpol.2018.05.063

Grigas, A. (2017). *The new geopolitics of natural gas*. Harvard University Press.

Grübler, A. (2012). Energy transitions research: Insights and cautionary tales. *Energy Policy*, *50*, 8–16. https://doi.org/10.1016/j.enpol.2012.02.070

Hardisty, P. E., Clark, T. S., & Hynes, R. G. (2012). Life cycle greenhouse gas emissions from electricity generation: A comparative analysis of Australian energy sources. *Energies*, *5*(4), 872–897. https://doi.org/10.3390/en5040872

Henriet, F., & Schubert, K. (2018). Is shale gas a good bridge to renewables? An application to Europe. *Environmental and Resource Economics*, *72*, 1–42. https://doi.org/10.1007/s10640-018-0223-2

Hondo, H. (2005). Life cycle GHG emission analysis of power generation systems: Japanese case. *Energy*, *30*(11–12), 2042–2056. https://doi.org/10.1016/j.energy.2004.07.020

IEA (International Energy Agency). (2018a). *Outlook for producer economies*. IEA Publications. www.iea.org/reports/outlook-for-producer-economies

IEA (International Energy Agency). (2018b). *World energy outlook 2018*. IEA Publications. www.iea.org/reports/world-energy-outlook-2018

IEA (International Energy Agency). (2021a). *Gas market report Q3–2021*. IEA Publications. www.iea.org/reports/gas-market-report-q3-2021

IEA (International Energy Agency). (2021b). *Net zero by 2050 – a roadmap for the global energy sector*. IEA Publications. https://iea.blob.core.windows.net/assets/ad0d4830-bd7e-47b6-838c-40d115733c13/NetZeroby2050-ARoadmapfortheGlobalEnergySector.pdf

IEA (International Energy Agency). (2021c). *World energy outlook 2021*. IEA Publications. www.iea.org/reports/world-energy-outlook-2021

IEA (International Energy Agency). (2023a). *Renewables 2023: Analysis and forecast to 2028*. IEA Publications. www.iea.org/reports/renewables-2023

IEA (International Energy Agency). (2023b). *World energy outlook 2023*. IEA Publications. www.iea.org/reports/world-energy-outlook-2023

IEA (International Energy Agency). (2024a). *Gas market report, Q1–2024*. IEA Publications. https://iea.blob.core.windows.net/assets/601bff14-5d9b-4fef-8ecc-d7b2e8e7449a/GasMarketReportQ12024.pdf

IEA (International Energy Agency). (2024b). *World energy investment 2024*. IEA Publications. https://iea.blob.core.windows.net/assets/60fcd1dd-d112-469b-87de-20d39227df3d/WorldEnergyInvestment2024.pdf

IGU (International Gas Union). (2015). *Life cycle assessment of LNG. 2012–2015 Triennium work report*. http://members.igu.org/old/IGU%20Events/wgc/wgc-2015/committee-reports-with-tnematic-sessions/pgcd-4-paper.pdf

IGU (International Gas Union). (2021a). *World LNG report 2021*. www.igu.org/resources/world-lng-report-2021/

IGU (International Gas Union). (2021b, October 28). *Getting it right to reduce methane emissions*. https://igu.org/news/getting-it-right-to-reduce-methane-emissions/

IGU (International Gas Union). (2024). *World LNG report 2024*. www.igu.org/resources/2024-world-lng-report/

IPCC (Intergovernmental Panel on Climate Change). (2001). *Climate change 2001: Synthesis report (third assessment report)*. www.un.org/en/climatechange/reports?gad_source=1&gclid=CjwKCAjwrIixBhBbEiwACEqDJXHRVdPbZzZJJhYEOIo06dSvzCo9dkIJzz7Ai1puNNIafErKCGMNORoCQgIQAvD_BwE

IRENA (International Renewable Energy Agency). (2018). *Hydrogen from renewable power: Technology outlook for the energy transition*. www.irena.org/publications/2018/sep/hydrogen-from-renewable-power

Jacobsson, S., & Johnson, A. (2000). The diffusion of renewable energy technology: An analytical framework and key issues for research. *Energy Policy*, *28*(9), 625–640. https://doi.org/10.1016/S0301-4215(00)00041-0

Kavalov, B., Petric, H., & Georgakaki, A. (2009). *Liquefied natural gas for Europe – some important issues for consideration*. Joint Research Centre of the European Commission. https://doi.org/10.2790/1045

Kemp, R. (2010). The Dutch energy transition approach. *International Economics and Economic Policy*, *7*(2), 291–316. https://doi.org/10.1007/s10368-010-0163-y

Kumar, S., Kwon, H. T., Choi, K. H., Lim, W., Cho, J. H., Tak, K., & Moon, I. (2011). LNG: An eco-friendly cryogenic fuel for sustainable development. *Applied Energy*, *88*(12), 4264–4273. https://doi.org/10.1016/j.apenergy.2011.06.035

Len, C., & Nian, V. (2014). Nuclear versus natural gas: An assessment on the drivers influencing Japan's energy future. *Energy Procedia*, *61*, 194–197. https://doi.org/10.1016/j.egypro.2014.11.1060

Lin, N., & Brooks, R. E. (2021). Global liquified natural gas trade under energy transition. *Energies*, *14*(20), 6617. https://doi.org/10.3390/en14206617

Mazzucato, M. (2018). *The entrepreneurial state: Debunking public vs. private sector myths*. Penguin.

McDowall, W. (2014). Exploring possible transition pathways for hydrogen energy: A hybrid approach using socio-technical scenarios and energy system modelling. *Futures*, *63*, 1–14. https://doi.org/10.1016/j.futures.2014.07.004

McGlade, C., Pye, S., Ekins, P., Bradshaw, M., & Watson, J. (2018). The future role of natural gas in the UK: A bridge to nowhere? *Energy Policy, 113*, 454–465. https://doi.org/10.1016/j.enpol.2017.11.022

McGlade, C., Speirs, J., & Sorrell, S. (2012). *A review of regional and global estimates of unconventional gas resources*. UK Energy Research Centre. https://d2e1qxpsswcpgz.cloudfront.net/uploads/2020/03/a-review-of-regional-and-global-estimates-of-unconventional-gas-resources.pdf

Meadowcroft, J. (2009). What about the politics? Sustainable development, transition management, and long term energy transitions. *Policy Sciences*, *42*(4), 323–340. https://doi.org/10.1007/s11077-009-9097-z

Moniz, E., Jacoby, H., Meggs, A., Kenderdine, M., & O'Sullivan, F. (2011). *The future of natural gas – MIT study*. http://energy.mit.edu/wp-content/uploads/2011/06/MITEI-The-Future-of-Natural-Gas.pdf

North, D. C., Wallis, J. J., & Weingast, B. R. (2009). *Violence and social orders: A conceptual framework for interpreting recorded human history*. Cambridge University Press. https://doi.org/10.1017/CBO9780511575839

Ouki, M. A. (2023). *Prospects for a potential African gas renaissance en route to a "just energy transition" OIES paper: NG, No. 185*. Oxford Institute of Energy Studies. https://hdl.handle.net/10419/280128

Pace Global. (2015). *LNG and coal life cycle assessment of greenhouse gas emissions*. Center for Liquefied Natural Gas. www.ourenergypolicy.org/wp-content/uploads/2015/10/PACE_Report.pdf

Perez, C., & Leach, T. M. (2018). Smart & green – a new "European way of life" as the path for growth, jobs and well-being. In *Council for research and technology development – re-thinking Europe. Positions on shaping an idea*. Council for Research and Technology Development. www.etis.ee/Portal/Publications/Display/cc517260-ed81-49ab-ab82-8651e3cfbaf7

Ritz, R. A. (2018). A strategic perspective on competition between pipeline gas and LNG, 2015. *The Energy Journal*, *40*(5), 195–220. www.jstor.org/stable/26756979

Roman-White, S. A., Littlefield, J. A., Fleury, K. G., Allen, D. T., Balcombe, P., Konschnik, K. E., Ewing, J., Ross, G. B., & George, F. (2021). LNG supply chains: A supplier-specific life-cycle assessment for improved emission accounting. *ACS Sustainable Chemistry and Engineering*, *9*(32), 10857–10867. https://doi.org/10.1021/acssuschemeng.1c03307

Roman-White, S. A., Rai, S., Littlefield, J., Cooney, G., & Skone, T. J. (2019). *Life cycle greenhouse gas perspective on exporting liquefied natural gas from the United States: 2019 update*. National Energy Technology Laboratory. www.osti.gov/biblio/1607677

Royal Dutch Shell. (2018). *Shell scenarios sky – meeting the goals of the Paris Agreement*. www.shell.com/news-and-insights/scenarios/what-are-the-previous-shell-scenarios.html

Schiffer, H.-W., & Trüby, J. (2018). A review of the German energy transition: Taking stock, looking ahead, and drawing conclusions for the Middle East and North Africa. *Energy Transitions*, *2*(1–2), 1–14. https://doi.org/10.1007/s41825-018-0010-2

Schuit, B. J., Maasakkers, J. D., Bijl, P., Mahapatra, G., van den Berg, A.-W., Pandey, S., Lorente, A., Borsdorff, T., Houweling, S., Varon, D. J., McKeever, J., Jervis, D., Girard, M., Irakulis-Loitxate, I., Gorroño, J., Guanter, L., Cusworth, D. H., & Aben, I. (2023). Automated detection and monitoring of methane super-emitters using satellite data. *Atmospheric Chemistry and Physics*, *23*(16), 9071–9098. https://doi.org/10.5194/acp-23-9071-2023

Selden, Z. (2020). Taming the bear: American liquified natural gas (LNG) exports and the negation of Russian influence in Europe. *Global Affairs*, *6*(2), 149–165. https://doi.org/10.1080/23340460.2020.1780460

Shayan, F. (2023). *The gas troika on the European gas market: Russia, Iran and Qatar. Vol. 1: 2008–2015*. De Gruyter. https://doi.org/10.1515/9783110981957

Simpa, P., Solomon, N. O., Adenekan, O. A., & Obasi, S. C. (2024). Sustainability and environmental impact in the LNG value chain: Current trends and future opportunities. *World Journal of Advanced Research and Reviews*, *22*(02), 581–601. https://doi.org/10.30574/wjarr.2024.22.2.1399

Smith, A., Stirling, A., & Berkhout, F. (2005). The governance of sustainable socio-technical transitions. *Research Policy*, *34*(10), 1491–1510. https://doi.org/10.1016/j.respol.2005.07.005

Stephenson, E., Doukas, A., & Shaw, K. (2012). Greenwashing gas: Might a "transition fuel" label legitimize carbon-intensive natural gas development? *Energy Policy*, *46*, 452–459. https://doi.org/10.1016/j.enpol.2012.04.010

Stern, J. (2017). *Challenges to the future of gas: Unburnable or unaffordable? (OIES paper NG 125)*. Oxford Institute for Energy Studies. www.oxfordenergy.org/wpcms/wp-content/uploads/2017/12/Challenges-to-the-Future-of-Gas-unburnable-or-unaffordable-NG-125.pdf

Stern, J. (2020). *Methane emissions from natural gas and LNG imports: An increasingly urgent issue for the future of gas in Europe. OIES paper NG 165*. Oxford Institute for Energy Studies. www.oxfordenergy.org/wpcms/wp-content/uploads/2020/11/Methane-Emissions-from-Natural-Gas-and-LNG-Imports-an-increasingly-urgent-issue-for-the-future-of-gas-in-Europe-NG-165.pdf

Steuer, C. (2019). *Outlook for competitive LNG supply. OIES paper ET06*. Oxford Institute for Energy Studies. https://doi.org/10.26889/9781784671310

Tamura, I., Tanaka, T., Kagajo, T., Kuwabara, S., Yoshioka, T., Nagata, T., Kurahashi, K., & Ishitani, H. (2001). Life cycle CO2 analysis of LNG and city gas. *Applied Energy*, *68*(3), 301–319. https://doi.org/10.1016/S0306-2619(00)00062-3

Turnheim, B., Berkhout, F., Geels, F., Hof, A., McMeekin, A., Nykvist, B., & Van Vuuren, D. (2015). Evaluating sustainability transitions pathways: Bridging analytical approaches to address governance challenges. *Global Environmental Change*, *35*, 239–253. https://doi.org/10.1016/j.gloenvcha.2015.08.010

Tusiani, M., & Shearer, G. (2016). *LNG: Fuel for a changing world: A nontechnical guide*. Pennwell Books.

Van De Graaf, T., Sovacool, B., Ghosh, A., Kern, F., & Klare, M. (2016). *The Palgrave handbook of the international political economy of energy*. Palgrave Macmillan.

Wettengel, J. (2020, May 15). *Unravelling the climate footprint of U.S. liquefied natural gas*. Clean Energy Wire. www.cleanenergywire.org/news/unravelling-climate-footprint-us-liquefied-natural-gas

Wood, D. A. (2024). The evolution of global liquefied natural gas supply chains: A review. In D. A. Wood & J. Cai (Eds.), *The fundamentals and sustainable advances in natural gas science and engineering* (pp. 23–55). Elsevier. https://doi.org/10.1016/B978-0-443-13420-3.00013-5

Wood Mackenzie. (2017, June 25). *Upstream carbon emissions: LNG vs pipeline gas [editorial]*. www.woodmac.com/news/editorial/lng-pipeline-gas-emissions/

Wood Mackenzie. (2020, August 17). *Renewable energy could reduce Asia Pacific LNG plant emissions by 8%* [Press release]. www.woodmac.com/press-releases/renewable-energy-could-reduce-asia-pacific-lng-plant-emissions-by-8/

3 The United Kingdom

Introduction

In the UK, a purposive energy transition is taking place towards a net-zero 2050 goal. Before transitioning to a gas-based system, the UK depended heavily on coal for energy. Today, with rapidly diminishing domestic reserves, decreasing production of oil and gas, and strong sociopolitical pressure to decarbonise, the country is transitioning towards low-carbon energy. The UK has an advanced integrated gas and electricity grid system, a mature nuclear power capability, a liquid market, and a thriving renewables sector. The country also enjoys unique potential for large-scale CCS, supported by a developed North Sea–based infrastructure and strong domestic energy industry. Policymakers have taken an "all of the above" approach to meet demand and address energy and climate policy goals. Economic development, security of supply, and market liberalisation all are issues that policymakers have tackled using an innovative mix of policy instruments.

Imported natural gas plays a key role in the current UK energy market, replacing dwindling domestic supply and meeting demand in key economic sectors. As carbon budget targets have become increasingly more challenging, especially insofar as they require decarbonisation in hard-to-abate sectors (e.g. heating and heavy industry), it has become crucial to analyse the role of LNG in the context of the overall future of the gas, hydrogen, and renewable energy mix. This chapter follows the T-E-T-I approach (introduced in Chapter 2) to provide in-depth insights into LNG – associated resources, market, energy-transition dynamics and power aspects – in the UK.

Natural gas and LNG in the UK

History and context

Throughout modern history, the UK has been a key player in the energy world as a producer, exporter, user, and now importer of various forms of energy. The country was the epicentre of the Industrial Revolution and the first to transition to fossil fuel use (Geels, 2011). The coal industry was a fundamental element of the country's energy system for nearly two centuries, including for heating, as a key supplier of feedstock to town (or artificial) gas, then as a competitor to natural gas.

DOI: 10.4324/9781003545149-3

The discoveries and development of the steam engine and later the high-pressure steam locomotive, coupled with other industrial applications, made coal the dominant source of heat and energy in the UK until the end of the nineteenth century. By 1900, coal accounted for 90% of the world's primary energy (Smith et al., 2005). Between 1877 and 1914, end-uses of gas in the UK were developed to increase access to the wider population. A state-led "technology substitution-type transition" took place between 1948 and 1977, with a move from coal-based to oil-based gas production (Fouquet, 2010). British coal became increasingly less competitive through lower investment in the sector, international competition, and trade union resistance to change led by the National Union of Mineworkers (Gouiffes, 2007). During that time, the UK gas industry was limited in its production and use of town gas as competition from imported oil increased and no new sources of gas were discovered. For a short period in the mid-1960s, Algerian LNG provided about 10% of the UK's total gas supply. This preceded a new natural gas socio-technical regime, triggered by the 1965 discovery of natural gas in the North Sea, 40 miles off the northeast coastal town of Grimsby, heralding the start of a series of discoveries of significant gas-rich fields (Gouiffes, 2007). Thus, the UK became an exporter of oil and gas, reaching peak production in the first decade of the 2000s. A sophisticated gas market and associated infrastructure were built during this time, alongside the development of a European gas market that began to evolve in the 1960s following natural gas discoveries in Groningen, a northern province of the Netherlands, and Western Siberia in the then Soviet Union.

LNG played an important role in the UK as an enabler in the coal-to-gas transition. In the late 1950s, a successful experiment to ship LNG to Canvey Island from Venezuela paved the road that went on to provide crucial backup during high-demand periods and support the development of the natural gas energy system (Gustafson, 2020). Following the Great Smog of London in 1952, the 1956 Clean Air Act created the conditions for the first commercial shipment of LNG to be delivered from Algeria to Canvey Island in 1964 (Arapostathis et al., 2013). LNG imports continued until the early 1980s when North Sea natural gas replaced imports. The 1990s were characterised by the "Dash for Gas", spurred by more efficient combined-cycle gas-turbine technology, which made the use of gas more attractive. Gas began to play an increasingly important role in UK life, with major consumers in power and residential heating (Bridge & Bradshaw, 2017; Corbeau & Ledesma, 2016).

Although global gas demand is forecast to increase in almost all scenarios in the coming decades, most of that growth is expected to occur in developing countries. The UK is one of several major economies with significant current gas demand that is forecast to decline in the short to medium term (HM Government, 2022, 2023; IEA, 2017, 2019). Nonetheless, natural gas will continue to be part of the energy mix in the short to medium term, where gas-on-gas competition takes place in a liquid, liberalised gas market supported by multiple LNG terminals and two gas pipelines (Bridge et al., 2013). With shale gas exploration and production unlikely to progress because of negative public opinion and limited resources (BP, 2019b; HM Government, 2022), LNG is forecast to play an important role to meet seasonal demand and for energy security reasons; it provides flexibility in an energy system that is moving to electricity produced from renewable sources and with declining natural gas production (BP, 2019a; HM Government, 2023). Longer term, natural gas's share in the energy system beyond the low-carbon bridging phase will depend on the UK's ability or political will to meet its decarbonisation targets while addressing its difficult-to-decarbonise sectors, primarily heating (Selden, 2020; Trutnevyte, 2016).

Competition, energy security, climate change, and decarbonisation are forces affecting energy policy and the trajectory of the UK's energy system. The UK has global, regional and, most recently, domestic, legally binding commitments to bring all GHG emissions to net-zero by 2050. The country has been successful in reducing emissions in the past, mainly because of the role natural gas played in replacing coal in power generation. In the long term, however, it is unclear what role natural gas and LNG will play in a decarbonising energy system that could yet undergo major transformation.

UK energy mix

The UK is the sixth largest economy in the world and the second largest in Europe, with a GDP of over US$3 trillion in 2023 (IMF, 2024). The structure of its energy system has shifted through the combined effects of the decarbonisation agenda, de-industrialisation, and growth of a consumer society. The country has experienced a rapid decline in production owing to declining reserves. Oil and natural gas production peaked in 1999 and 2000, with 3 million barrels and 3.8 TCF per day, respectively. By 2017, total energy production had declined 57% from its 1999 peak, with significant drops in oil and gas production, preceded by an even steeper reduction in coal production since the 1970s (IEA, 2019).

Natural gas production has remained steady since 2014, after a long period of decline. Despite the potential for shale gas, environmental concerns mean that the industry is unlikely to be developed, and natural gas production is forecast to decrease steadily over the coming decades because of diminishing reserves. Coal and natural gas have historically competed for a share of the electricity market, with coal's overall share declining rapidly, initially at the expense of natural gas but increasingly to renewables. Following a peak in consumption in 2004, the consumption of natural gas has steadily decreased across all sectors in the UK, initially associated with increased domestic energy efficiency and insulation, leading to reduced heating demand, and later to higher levels of renewable energy power generation. In 2017, the UK was at the median level of the IEA's member countries in terms of fossil fuels' share of total primary energy supply (IEA, 2019). In 2022, 39% of the UK's primary energy consumption was of natural gas, demonstrating the importance of natural gas in the country's energy mix. Oil constituted 35% of the country's consumption and renewables 17% (IEA, 2023; National Grid, 2023). Focusing on natural gas demand by sector, the residential market comprised the largest sector (40%), followed by the power (26%) and industrial (16%) sectors.

Carbon dioxide emissions

Between 1990 and 2022, the UK's GDP grew by 60%, while its CO_2 emissions reduced by 50%, reaching their lowest levels since the late 1800s (Department for Energy Security and Net Zero, 2024a; ONS, 2024). The UK managed to decouple CO_2 emissions from economic growth by reducing the carbon intensity of its energy supply and the energy intensity of the economy (IEA, 2018). Decreases in CO_2 intensity took place faster in the UK than in other IEA countries mainly because of the rapid increase of renewables, particularly in power generation. In 2018, energy-related emissions, led by oil and natural gas, accounted for approximately 77% of total GHG emissions, followed by the agricultural, industrial, and water sectors (National Grid, 2019b). The Department for Business, Energy and Industrial Strategy (BEIS)[1] projects that by 2030 emissions will be 50% below 1990 levels, given continued reduction in electricity-related CO_2 emissions, an increase in nuclear power generation, and the closure of coal-powered plants (IEA, 2018). The UK achieved success in reducing emissions mainly because of the role natural gas played in replacing coal in power generation.

Natural gas dynamics and infrastructure

The UK became a net importer of natural gas in 2004, and by 2022 it imported 2.0 TCF of gas, 55% of which was through pipelines and 45% by LNG (Department for Energy Security and Net Zero, 2023a; EIA, 2024). The National Transmission System includes approximately 7,600 km of pipelines, with three direct gas pipelines connecting to other countries (IEA, 2018). Import pipelines connected to Norway and continental Europe have an import capacity of 100 billion cubic metres (BCM) per year (IEA, 2019). The UK also has significant export capability, with 23 BCM exported to continental Europe and Ireland in 2022 (Department for Energy Security and Net Zero, 2023b). Storage capacity in the country is low, with 1.4 BCM in total, representing approximately 2% of total demand in 2017. Centrica Plc decided to end operation of its Rough gas-storage facility in 2017, the last long-range gas-storage facility in the UK, owing to lower seasonal spread prices and high maintenance and operation costs (IEA, 2019). In October 2022, Rough reopened at 20% the capacity of its previous capacity for 2022/2023 winter bolstering storage. Rough now provides approximately half of the UK's total gas storage.

The UK increased its gas imports through interconnector pipelines and LNG terminals from 2002, with an increase of 40% between 2007 and 2022 to 54.2 BCM, with the United States, Qatar, and Norway accounting for 90% of total UK imports in 2022 (EIA, 2018, 2024; National Grid, 2023; Toke, 2013). Import dependency was approximately 40% in 2018 and has increased gradually as domestic production has reduced (Pye et al., 2019). LNG import capacity is approximately 48 BCM/year, with three operating LNG terminals (Department for Energy Security and Net Zero, 2023b). In 2022, interconnecting pipelines and LNG import terminals were importing well below their capacity (at approximately 30% and 50%, respectively) (National Grid, 2019a, 2023).

Natural gas forecasts

Coal, nuclear, and renewables

BEIS (2018) predicted that coal will comprise less than 2% of all fuel demand by 2030, none of which will be for power generation. The low-cost benefits of coal-to-gas switching have been realised in the UK with the support of a carbon floor price, but they do not address the role of the gas industry in a decarbonised system (CCC, 2019).

The construction of additional nuclear power plants has been debated since the 1980s based on wavering public support and the industry's debts, largely based on the commercial and technical challenges of the Sizewell B nuclear power plant. Although the role of nuclear energy is controversial with a complex history, it continues to play an important role in meeting low-carbon power generation needs and retains strategic importance as an industry. Its future, however, will largely depend on whether political support is sufficiently strong to ensure market intervention, especially in the context of rapidly declining renewable energy costs (National Grid, 2019a).

The future of renewables is affected less by technological uncertainties, and, thus, their long-term role in low-carbon electricity generation is more robust. Renewables are projected to be the main source of power generation for decades to come in several scenarios (CCC, 2019). The growth of renewables will play an important role in the future of natural gas as a fuel source, particularly for LNG, and consideration will need to be given to the intermittent nature of solar and wind sources and the UK gas market's seasonality, especially in the context of limited gas-storage facilities.

Natural gas and LNG

The role of natural gas in the future of the UK's energy system is less clear than for coal and oil, renewables, and even nuclear, with increasing criticism levied against the bridge-fuel argument, and CO_2 and CH_4 emissions coming under greater scrutiny in Europe and the UK than in other parts of the world. BEIS models project a gradual decline for gas-fired power generation up to 2035 when fossil fuels are displaced by nuclear and renewables (BEIS, 2017, 2018). The role and timing of natural gas as a transition fuel in the UK rest heavily on the industry's ability to successfully deploy CCS, and meet 2050 carbon targets, which will constrain natural gas's role beyond the late 2020s (McGlade et al., 2014).

National Grid's (2022) future energy scenarios range from the least ambitious and slowest credible decarbonisation scenario of "falling short" to the more ambitious "leading the way" scenario. The less-ambitious scenarios project less dramatic improvements in the decarbonisation of heat and more modest reductions in gas demand, with a focus on energy efficiency. In the more ambitious 2050 scenarios, more significant reduction in gas demand is projected, with greater emphasis on electrification and overall decarbonisation of heating. National Grid (2022) projects different amounts of gas imports and conventional United Kingdom

Continental Shelf (UKCS) production, with variability in the amount of hydrogen produced from natural gas with carbon capture, utilisation and storage (CCUS), shale and green gas connected to the distribution and transmission networks. In all scenarios, longer-term UKCS production reduces rapidly but is partly replaced either by green gas or domestic shale gas. Gas demand remains higher in scenarios where natural gas is used to produce hydrogen or for electricity generation with CCUS. Industrial gas demand is steady in some scenarios, with a more dramatic decline post-2030 except where natural gas is required for hydrogen production. In the net-zero 2050 scenarios, natural gas's role in residential demand is mainly reduced to the supply to produce hydrogen to be used as part of the roll-out due to begin between 2030 and 2035 (National Grid, 2022).

In terms of the natural gas profile, with domestic production decreasing and the production of unconventional gas being unlikely, imported gas will play an increasingly important role as a share of total gas used. There are few prospects of increasing gas volume import capacity from Norway and continental Europe if no additional pipeline capacity is built or new reserves discovered. This means that LNG imports from approximately 6 BCM in 2018 up to 12 BCM in 2050 will be required to fill that gap, with the actual volume dependent on the decarbonisation approach taken.

Key drivers and tensions

Key themes identified as affecting the role of LNG in the future of the UK's energy system were extracted from interviews with industry role-players (for further details of the interview participants and process, see Al-Kuwari, 2023). These themes included climate change and "leading by example"; decarbonisation and net-zero 2050; North Sea infrastructure and the potential for hydrogen and CCS; security of supply and competitiveness.

Climate change and "leading by example"

The UK has a history of progressive environmental policies and, more recently, climate-related policies that will affect the trajectory of natural gas as a fossil fuel. After decades of energy policy driven by competing (but sometimes complementary) goals, the environmental agenda was prioritised towards the end of the twentieth century. The Climate Change Levy was introduced in 2001 as a charge on the supply of energy to public-sector customers and businesses, providing different rates per unit of energy for electricity, gas, coal, and liquified petroleum gas (IEA, 2019), while the 2003 Energy White Paper set out the future of a low-carbon economy (Department of Trade and Industry, 2003). The political importance of climate change was generally low up to 2006, with the UK on track to meet its Kyoto Protocol targets and little overall pressure to take on more aggressive measures (Carter & Jacobs, 2014). In 2008, however, the UK's long-term climate policy was developed under the Climate Change Act, which provided a legal framework for reducing and mitigating GHG emissions and adapting to the impacts of climate change. This was positioned by the Government at the time as an answer to both climate change and energy security objectives. By 2013, with the public's focus on the ongoing recession, the Government and Treasury took advantage of cost concerns and slowed down climate policy momentum, putting economic development ahead of green objectives (Geels et al., 2016).

Internationally, the UK took on a strategic role as part of the European Union (EU)'s Troika Presidency in the Kyoto Protocol negotiations and continued to position itself as a leader in action to mitigate climate change. Following formal ratification of the Paris Agreement as an EU Member State in 2015 (UNFCCC, 2015), the UK's Committee on Climate Change (CCC) recommended more aggressive domestic reduction targets, based on capability, equity, and supporting the global effort. In May 2019, a target was set for net-zero GHG emissions by 2050 (CCC, 2019).

In previous decades, climate advocates strategically tied concerns regarding foreign dependence with climate goals as a means of promoting domestic renewable energy production and to push for more aggressive climate reforms (Kern et al., 2014). UK policymakers could leverage the increased popularity of the climate change agenda and continue to use climate policies to meet other objectives, such as increasing self-sufficiency and addressing security of supply concerns by limiting the overall use of fossil fuel imports and supporting decarbonisation. A public attitudes survey (BEIS, 2019c, 2023) concluded that addressing changing behaviours and climate change, reducing the carbon footprint through transportation, installing household low-carbon heating systems, and minimising energy at home were the behaviours most often cited as affecting daily life for UK citizens, highlighting the importance of climate change in daily life.

The UK has signed up to major protocols and set itself more ambitious targets than many of its European neighbours. Al-Kuwari's (2023) interviews undertaken with British stakeholders provided insights into the national psyche – that once the British public decides something is morally correct or ethically driven, then the focus shifts to that mission regardless of any divisions around policy methods. This ties into the "leading by example" approach, which plays a role in maintaining public support for the net-zero target. Broad support exists, with 86% of those surveyed agreeing with the statement: "If everyone does their bit, we can reduce the effects of climate change". This highlights the importance of the link between domestic and international commitments and overall public support (BEIS, 2019c). However, a portion of the population, particularly older generations, may be sceptical of why the UK is going above and beyond other countries when the overall impact of UK efforts on global emissions will be relatively inconsequential from a net GHG emissions perspective. One interviewee summarised this view as "Why are we doing this when . . . the Chinese are going to build another 500 coal-fired power stations [in] the next ten years?" (Al-Kuwari, 2023). This dynamic may be a factor in whether policies deemed to be climate related become more economically challenging for future governments.

The British media strongly influence energy and climate social-related priorities domestically. One interviewee stated that the debate has been shaped by a narrative of "Investment in oil and gas is bad, while investment in wind is good" (Al-Kuwari, 2023). This position has added to an increasingly polarised environment, which has affected the ability to have a rational public discourse on the dynamics affecting potential future policy pathways. While the UK has been among the global leaders in addressing climate issues, it is also heavily dependent on natural gas. A combination of a weak gas lobby and the EU-wide movement from civil society and the media may push EU politicians to take a more aggressive stance towards natural gas, potentially even banning it completely by the end of this decade. This would put additional pressure on the UK to manage its energy system and coordinate with European markets. Natural gas and LNG face challenges because the industry has failed to maintain the transition fuel narrative in the fight against climate change. As Stern (2019) concluded, the gas community in the EU can no longer claim its product to be low carbon given the increased levels of scrutiny on CH_4 emissions. The "leading by example" approach means the debate around natural gas risks has become increasingly polarised, which could adversely affect the overall role of natural gas, turning it into just another fossil fuel.

Decarbonisation and net-zero 2050

"Not in my back yard!"

Despite general enthusiasm for decarbonisation, planning and development of new projects may face local challenges from a NIMBY-ism mentality ("Not in My Back Yard!") and an increasingly litigious society. The Government could find it challenging to implement major infrastructure changes, such as retrofitting the gas grid or increasing onshore wind capacity, because of residents increasingly opposing developments in their local area. The failure to develop shale gas in the UK is partly attributable to this. Opposition to fracking has increased, from 21% to 36% from 2013 to 2022 (down from 45% in 2021) according to the BEIS (2023) attitudes tracker, with the most common reason being the destruction of nature, the need to focus on renewable energy development, and the risk of contamination and earthquakes. Through crowdfunding and other means, citizens can raise funds for causes on an unprecedented scale, with the resulting legal challenges hampering initiatives.

Heating and hard-to-abate sectors

Decarbonisation of the heating sector requires significant changes and will determine in large part the future role of natural gas. Heating contributes approximately one-quarter of all GHG emissions in the UK, and the sector must be nearly fully decarbonised by 2050 to meet targets and goals linked to the 2008 Climate Change Act and Paris Agreement (CCC, 2018, 2019; HM Government, 2022; National Grid, 2023; Staffell et al., 2018). Successful transition to a decarbonised domestic heating system will require public awareness and societal involvement because some pathways could lead to major changes in a well-established infrastructure system (CCC, 2019).

The Clean Growth Strategy sets out two main potential pathways to decarbonising the heating sector – an electricity-based path focused on electrifying the UK's heat demand, primarily using low-carbon electricity-powered heat pumps; and a gas-based path, supported by hydrogen produced initially from natural gas with carbon capture and then by low-carbon electrolysis. Both pathways require significant development and investment to reach the scale needed for 2050, including the development of CCS, and a significant increase in low-carbon energy capacity (IEA, 2019). Studies and policy recommendations by National Grid, the UK Energy Research Centre (UKERC), and the CCC have laid out a series of "low or no regrets" potential actions to address uncertainties related to heating policy. The near-term measures are focused on improving thermal efficiency in homes and rolling out installation of low-carbon heat pumps. Longer term, the displacement of natural gas will be challenging given its familiarity and convenience for end-users. Decisions centred around the difficulty of fully replacing natural gas will be crucial in determining the roles of hydrogen and electrification in a future energy system. Short-term initiatives can help to facilitate measures beyond 2025, including decisions on heat decarbonisation for new buildings (CCC, 2019; HM Government, 2022, 2023; National Grid, 2019a, 2023; Whinskel et al., 2019).

More focused and ambitious policies may be needed to address the hard-to-abate sectors. In terms of residential heating, there is yet to be an effective implementable decarbonisation pathway, especially one that does not include natural gas. In addition, the narrative on decarbonisation of heating should progress from discussing full electrification to include molecules, specifically hydrogen, recognising that there needs to be a role for gases in the absence of viable alternatives in the short to medium term.

Decarbonising heating and transport require decisions that will affect many consumers, and lifestyle considerations that must be accounted for. There may be significant potential pushback from consumers owing to the convenience of natural gas and the amounts already invested in domestic boilers and other gas appliances. A pathway involving natural gas blended with hydrogen is potentially a more favourable, less intrusive route. However, nearly 20 million households would still require adjustments to appliances and boilers – a complex challenge that will require considerable thought from the Government.

Renewables

Electrification is the major tool for decarbonising the UK energy system. In National Grid's (2019a) 2°C scenario, electricity consumption will increase by nearly 50% by 2050 from 2018 levels. Natural gas's flexibility means that it will continue to have an important role during the transition, even in the power sector. Natural gas is the only practical alternative to support large-scale renewables until the technological and economic challenges of deploying low-carbon storage at scale are solved. Thus, the intermittent nature of renewables establishes a stable role for natural gas in "balancing the market" in short-term and seasonal perspectives. The development of the hydrogen industry and blue and green hydrogen pathways, and links to large-scale renewable generation (primarily offshore wind) will play a key role in determining the scale and timescale of the threat to natural gas, that is, the point at which hydrogen can shift from being an opportunity for natural-gas-produced blue hydrogen to becoming a threat to electrolysis-produced hydrogen.

Net-zero, not zero

Decarbonisation is a key driver in the UK's energy transition and a principal factor in determining the future of natural gas and LNG. Decarbonisation became a driver partly because of the increased awareness of climate change; however, it is also affected by and related to other factors, including security of supply, economic development, and the UK's future industrial policy. In the long term, decarbonisation can be achieved through several pathways, all of which will affect import dependency, competitiveness, and emissions. In the short to medium term, pivotal decisions need to be taken to overhaul the energy system, requiring significant economic resources. Decarbonisation and the green economy, followed by energy security, affordability, and industrial strategy, were the main energy policy drivers identified in the UKERC's 2019 expert survey (Whinskel et al., 2019).

BEIS (2017) proposed a Clean Growth Strategy, part of the UK Government's industrial strategy that set out the decarbonisation agenda for the UK economy through to 2032. The proposal combined a series of interventions and instruments targeting energy regimes with sectoral objectives to support new pathway development (IEA, 2019). One UK policymaker described clarity on the way forward with no visible potential pressure to divert away from the low-carbon vision (Al-Kuwari, 2023). Stakeholders from the energy industry, however, were sceptical about the UK's ability to achieve ambitious long-term decarbonisation targets, believing that there is currently no clear framework by which current governments and politicians can be held accountable for targets beyond their political term, despite the existence of a net-zero 2050 goal and a five-year carbon budget mechanism. To meet net-zero by 2050, the electricity system will need to operate under lower-carbon generation production mechanisms complemented by new technologies, such as negative emissions in the power sector and potentially the use of hydrogen fuel for several applications, including heating. National Grid outlined multiple pathways to reach the 2050 target, each with trade-offs to be considered. Multiple sociopolitical uncertainties exist beyond the technical challenges, including large-scale storage, carbon capture at scale, and aggressive sectoral policies.

Li and Pye (2018) interviewed a diverse group of participants about the UK's ability to meet its net-zero target. Sociopolitical dimensions of uncertainty were mentioned almost as often as technological dimensions, with differing views on the role of government and society in the transition – similar findings to Al-Kuwari (2023). From a policy development perspective, strong regulatory and legal frameworks are already in place and new initiatives can build on the strengths of the system. In the past, the UK has succeeded in reducing its emissions – a success on which to build. Even though energy intensity has decreased consistently through effective policymaking, overall economic underperformance, coupled with a shift towards a service-based economy, makes it difficult to assess whether that success in reducing emissions can be replicated in the future.

Despite the UK's departure from the EU, the net-zero target will be supported by the EU's targets and overall climate policies will remain broadly aligned. Links to Europe may provide the UK with momentum from a regulatory and policy framework perspective that is conducive to decarbonisation. There is potential for political and climate pressure to push the EU to adopt a ban on all fossil fuels as early as 2030, by "flipping the switch", in which case the UK's net-zero targets would be more likely to be achieved. In terms of LNG imports, new trade agreements may create complications in relation to overall emissions and specifically CH_4 emissions. What happens in the United States – the world's largest LNG exporter in 2023 – is of particular interest in relation to how these agreements will affect the economics of future supplies.

Opportunities exist for exporting knowledge, skills, and competencies, especially in CCS, and hydrogen and offshore wind, presented by a net-zero agenda. This view is countered, however, by concerns about the Government's ability to meet competing economic pressures, and its willingness to risk delivering on basic priorities given potentially unclear future financial benefits. To meet the UK's overall emissions-reduction commitments, the 2008 Climate Change Act introduced carbon budgets to provide a five-year, statutory cap on total GHG emissions. The UK successfully met the first, second, and third carbon budgets (Department for Energy Security and Net Zero, 2024b); however, there are projected shortfalls for the fourth and fifth budgets, placing its legally binding commitments at risk (BEIS, 2017; CarbonBrief, 2023). The fifth carbon budget, adopted in 2016 to cover the 2028–2032 period, targets a 57% reduction in GHG emissions relative to 1990 levels (IEA, 2019). In its analysis of the UK's progress towards net-zero, the CCC pointed to a massive policy gap, with insufficient funding and underperformance of current policies in areas such as transportation, CCUS, and nuclear energy. Key challenges in meeting the targets also include decarbonising the power sector, delivering affordable energy, and establishing a post-Brexit framework. Watson et al.'s (2015) survey of experts identified factors likely to determine the UK's ability to meet the fourth carbon budget, including access to finance for low-carbon generation, the availability of commercialisation initiatives for low-carbon power-generation technologies, including support for CCS, and stronger measures for energy efficiency improvements.

All decarbonisation options, and not just electrification, should be explored to ensure that the most technologically and economically viable pathways are followed. In approaching net-zero and its impact on natural gas, the ambition should be net-zero, not zero, allowing for flexibility in determining different pathways and for natural gas to potentially play a positive role in the UK energy system's evolution.

North Sea and the potential for hydrogen and CCS

As UK oil and gas production decreases, the North Sea will continue to host several depleted fields and a well-developed offshore pipeline network that could be used for several purposes, including CCS. Offshore depleted fields are valuable because their use for storage is likely to encounter less public resistance than the development of new onshore storage. Decommissioning older oil and gas fields is forecast to cost £1.5 billion per year to 2027 (Hinson et al., 2020), and the industry would have an incentive to explore ways to minimise costs. The North Sea has a potential role in determining the future trajectory of the energy system to support large-scale electrification and underpin hydrogen production; this could be achieved through retrofitting the existing natural gas infrastructure to produce and transport hydrogen, developing carbon storage locations, and expanding offshore wind power. Current infrastructure is a key competitive advantage for the UK in terms of supporting the acceleration of hydrogen use and demand. Furthermore, the North Sea, and offshore UK locations in general, are key to the UK's successful development of a world-class wind industry in a short period of time. Overall, the UK's long history of using the North Sea for its energy needs creates expertise that can be leveraged to make it a global player as a developer, funder, and operator of complex projects.

The potential for natural-gas-derived hydrogen, coupled with CCS, will play a significant role in determining LNG's potential in the future UK energy mix. CCS and hydrogen are two different, but linked, technologies; one pathway for decarbonising natural gas involves the production of hydrogen using natural gas coupled with CCS through steam CH_4 reforming (SMR) and autothermal reforming (ATR) methods.

Without commercial scale-up of CCS, natural gas's role could be very limited in a net-zero energy system. Pye et al. (2017) concluded that unless demand for energy drops by unprecedented levels or major CCS technology breakthroughs occur, a 2050 net-zero system is unlikely, highlighting the importance of CCS to the UK. Although the success of renewable gas or green hydrogen is independent of CCS availability in the long term, it would benefit from the increasing demand created by blue hydrogen before green hydrogen can become competitive.

Carbon capture and sequestration

The UK's offshore geology provides clear opportunities for CO_2 storage, with potential for storing 78,000 million tonnes at over 500 potential sites (HM Government, 2023). These sites are mostly located near industrial clusters (IEA, 2019), providing some of the infrastructure needed to scale up CH_4 reforming combined with CCS. The UK, along with China, Norway, and Canada, has plans to develop large-scale CCS projects to meet its climate goals (HM Government, 2023; IEA, 2018). Approximately 63% of those surveyed by the BEIS (2023) stated that they were aware of CCS, with 44% indicating general support for it as a technology.

The UK supported CCS early on, with a 2012 road map towards its commercialisation by 2020. The objective was to deploy over £1 billion of capital funding to build knowledge to reduce future costs and facilitate the development of a stable and appropriate regulatory framework (CCC, 2016). Following a bidding process leading to the selection of two projects and approval of nearly £100 million in R&D funding, the initiative was cancelled in 2015, a few weeks before final bids were due (Krug & Lebelhuber, 2018). The late-stage cancellation of the CCS competition indicated a view within government that CCS was too expensive and lacked feasibility, calling into question future pathways for CCS development in the UK. Despite these setbacks, other support measures were established to support CCS. The CCUS action plan also laid out plans for CCUS deployment. These measures are defined as triple-lock policies – carbon-capture readiness as a condition for coal-fired power station planning consent; a carbon price floor topping up the EU emissions trading scheme (ETS); and an emissions performance standard requiring new coal-fired plants to be equipped with CCS (IEA, 2019; Krug & Lebelhuber, 2018).

Interviewees expressed scepticism about the UK Government's ability to support the development of CCS at scale, pointing primarily to the lack of consistent government funding and commitment (Al-Kuwari, 2023). One industry insider noted: "The Government needs to re-establish its credibility in that field after its very chequered policy past where it sort of withdrew support for CCS" (Al-Kuwari, 2023). From a financing perspective, there are measured expectations about the ability to deploy large-scale CCS, with one interviewee concluding that it is unlikely that it will be the "silver bullet it was commonly seen as ten years ago". Moreover, climate activists in the UK have warned that CCS must be developed with caution because it could delay the transition away from fossil fuels (Pye et al., 2019). Some social and academic stakeholders, policymakers, and the non-fossil fuel community are concerned by the moral element of the net-zero debate. These "hydrocarbon rejectionists", in Stern's (2019) words, do not accept any form of fossil fuel, running counter to the position taken by many energy modellers who envisage a phasing-out period for fossil fuels based on techno-economics (McGlade et al., 2018). The potential role of CCS in facilitating pathways towards net-zero 2050 in the UK has been fundamental in shaping the debate, laying out conditions for decarbonisation and providing a way forward for natural gas to be recategorised as low-carbon hydrogen, potentially making it part of the solution, not the problem.

Hydrogen

Although hydrogen is currently a key focus for UK energy policy and the UK is one of only six countries with a national hydrogen and fuel cell target (Staffell et al., 2018), hydrogen's long-term role in the UK's energy system remains uncertain. Stakeholder opinions on its role differ widely, with some viewing it as highly disruptive and others as reinforcing current industrial and commercial patterns (Al-Kuwari, 2023; McDowall, 2014). Decarbonising the current natural gas grid by using hydrogen as a replacement gas leverages current infrastructure and reduces disruption to the gas-dependent energy system.

There is precedent for such mass conversion in the UK – namely, the shift from town gas to natural gas in the 1960s – but a change from natural gas to hydrogen would be more complex, from both technical and regulatory perspectives, partly as a result of the privatisation of networks (Dodds & Demoullin, 2013). National Grid launched the Hydrogen in the National Transmission System (HyNTS) programme with the aim of identifying the challenges and opportunities involved in transporting hydrogen in a way that would be consistent with the UK's 2050 net-zero targets (CCC, 2016). Natural gas reforming with CCS is currently the lowest-cost option for producing hydrogen because the levelised cost of hydrogen for electrolysis is approximately double that of SMR or ATR owing to a combination of electricity prices, efficiency levels, and unit costs (HM Government, 2021; National Grid, 2019a). There are no significant cost reductions expected by 2050 for reforming technologies, providing an opportunity for new technologies to play a greater role. Electrolysis-produced green hydrogen requires enormous amounts of surplus low-carbon electricity, making it cost-prohibitive today and unlikely to be used at scale in the UK in the near term (National Grid, 2019a).

The CCC explored different hydrogen supply scenarios and made a series of recommendations. In the "niche" and "hybrid" scenario approaches, where hydrogen would initially target hard-to-abate sectors and gradually other applications (CCC, 2019), CH_4 reforming coupled with CCS would be deployed initially to become an established solution and capture scale until CH_4 produced via electrolysis and renewable energy develops at scale and improves efficiencies. In the two "full hydrogen" scenarios, over 90 GW of hydrogen capacity would be installed over a 20-year period, one using a primarily natural-gas-based hydrogen and another a low-carbon variant. In the "natural gas" scenario, 140 $MtCO_2$ per year would be emitted by 2050 in hydrogen production alone, many times the CCC's recommended 20 $MtCO_2$ per year by 2035, in addition to contributing to increased gas demand and imports. This highlights the carbon intensity of the reforming process, which is a function of the production of fossil fuels, incomplete CO_2 capturing, and residual emissions. In the "primarily electrolysis" scenario, more than 300 TWh of additional zero-carbon electricity generation would be required by 2050 (CCC, 2019), which equates to another 31 GW of nuclear capacity over the niche scenario, and consequently over 100 GW of electrolyser capacity to manage system and renewable energy imbalances and overall efficiencies – challenging and unrealistic low-carbon electricity generation capacity requirements.

The implications for hydrogen on natural gas were analysed in the CCC's (2019) hydrogen scenarios and National Grid's (2019a) future demand scenarios. To observe the 2°C scenario, CH_4 reform is the main choice to achieve production at scale, with electricity for electrolysis playing a relatively minor role. In future forecasts, assumptions around carbon prices and policies are central to the development of hydrogen fuel.

Hydrogen will be a key factor in shaping the future of natural gas in the UK, and the eventual direction taken will be pivotal to the future of LNG. However, the economics and technical feasibility that will determine that point remain uncertain. While countries such as Germany develop their green hydrogen economies – with plans focused on the prioritisation of green hydrogen in shipping, aviation, steel, and other chemical industries (German Federal Government, 2024; Pingkuo & Xue, 2022) – and policy decisions focused on achieving scale and reducing costs materialise in the EU and other regions, there is potential for green hydrogen to become more competitive than blue hydrogen. The hydrogen and fuel cell–based knowledge in the UK could also become another export-worthy knowledge base. The coupling of nuclear energy with electrolysis, with its potential for greater energy density, could provide a third route for low-carbon hydrogen production that could alter development plans.

In the UK, natural-gas-based blue hydrogen coupled with CCS has a potentially significant role to play in the energy industry, with green hydrogen as its long-term competitor (HM Government, 2021). Policy approaches could be developed for specific sectors, for instance, green hydrogen could be mandated for hard-to-abate sectors, which would limit the widespread adoption of hydrogen but still contribute to the reduction of carbon emissions in polluting sectors. In scenarios where hydrogen is more widely adopted, however, most of the gas needed for hydrogen production would have to be imported from LNG sources. Overall, the potential of CCS means that LNG in the UK could be a transition fuel to a hydrogen economy. The length of that transition will largely depend on the economics of green hydrogen and the successful scale-up of CCS.

Industrial clusters

Industrial clusters are areas that contain several energy-intensive industries, such as chemicals, glass, paper, oil refining, cement, and ceramics. These clusters are key economic hubs in the UK with an economic value of around £150 billion and creating 1.5 million jobs and over £300 million of exports in goods and services (BEIS, 2017). Policymakers have identified large industrial clusters with hydrogen supply networks as potential cornerstones of the early-stage hydrogen scale-up strategy to address hard-to-abate sectors. This hydrogen-focused approach aims to transform industrial clusters into low-carbon manufacturing hubs to support long-term emissions-reduction targets and increase inward investment (BEIS, 2017). One showcase scheme in this programme is the H21 Leeds City Gate Project. Leeds was selected based on its manageable size, well-developed infrastructure, and location near salt beds north of Hull and Teesside in north-east England. The project targets full hydrogen conversion of the gas system at a cost of approximately £2 billion, financed by Ofgem's innovation scheme (Krug & Lebelhuber, 2018). The H21 Leeds project is studying the potential to increase hydrogen use by converting existing natural gas networks and provide heat through a 100% hydrogen gas network to over 3.7 million homes and industries (Sadler et al., 2016). A "well to sofa" concept design begins with the extraction and processing of natural gas, transporting it to Teesside where it is converted using a CH_4 reforming plant, with the generated CO_2 transported and stored in the North Sea. Around

1.5 million tonnes of CO_2 per year would be sequestered, capturing 90% of the CO_2 emitted from four steam CH_4 reformers. The hydrogen is then stored in salt caverns to meet intra-day and inter-seasonal demand. Over three years, the distribution network will undergo a full conversion ensuring that all appliances are replaced with hydrogen-burning devices at no cost to home and business owners (Sadler et al., 2016).

If successful, the H21 Leeds City Gate project could provide the blueprint for other cities across the UK and Europe. The project showcases the technical characteristics that make the UK a dynamic environment for energy transition, bringing together technical variables that can be integrated and potentially replicated on an already significant scale. It also highlights a potential role for LNG as a key player in the path to decarbonisation by focusing on importation facilities located near industrial clusters. Such clusters provide the right platform for the goals and technologies identified to be tested and scaled up. There is an opportunity to increase coordination in hydrogen development across Europe by leveraging industrial clusters, especially within the Netherlands and Germany where a common grid could enable transportation and trade.

Security of supply

Security of supply has been a critical factor influencing energy policymaking in the UK for decades. Concerns around importing sources coincided with the UK's shift from its previous exporter role of oil and gas to becoming an overall importer. This amplified the public and political debate about the UK's energy dependence, and possibly paved the way for a public refocus towards domestic production to minimise exposure to energy imports. There is also particular focus on natural gas in the EU's energy policy, with energy dependence being a fundamental political union tool for the block (Offenberg, 2016). The Russia/Ukraine war has greatly increased scrutiny of the reliability of foreign imports into Europe (HM Government, 2023); Kern et al., 2014). The 2007 UK Government's "Meeting the Energy Challenge" White Paper highlighted economics and energy security as key issues, contributing to the emergence of a new narrative that combined geopolitics, energy security, and climate change (Geels et al., 2016; Kern et al., 2014).

The public's overall concerns about future energy security have remained stable or decreased slightly over time, with price increases and the UK's pace of investment in alternative energy sources and over-dependence on energy imports being the main focus areas (BEIS, 2019a, 2019b, 2019c, 2023). The BEIS surveys highlight the difficult trade-offs for UK policymakers during the energy transition, where concerns around price increases and an over-reliance on energy imports must coexist until sufficient domestic sources of energy become available.

Although climate change and environmental issues have gained prominence in the energy policy discourse, energy security is a consistent driver for UK policymakers. A review of UK government statements on energy policy by Toke and Vezirgiannidou (2013) found that the main lens through which energy security is portrayed projects the Government's policy objective at that time. Other countries have used security of supply to protect their domestic industries, as Germany and China have done with their domestic coal industries. Security of natural gas supply in the UK is underpinned by a large liquid market, a developed infrastructure, and a diversified mix of supply by geography and type. Flexibility is supported by the market's ability to send effective supply- and demand-driven price signals that allow gas to flow back to the UK to cover any shortages, underpinned by significant interconnector and LNG import capacity (IEA, 2019). The interrelatedness of the gas and electricity markets means that the system can take advantage of strong wind outputs and the interconnectedness of European and UK markets, all with the possibility of attracting diverted LNG cargoes as and when needed, thus stabilising the market and the supply-and-demand profile.

The importance of security of supply has also increased focus on storage, where LNG terminals represent additional potential gas supply and play a part in replacing traditional storage (e.g. the recently closed Rough storage facility). The recent increase in the number of importers – from 3 to 16 in 2018 alone – highlights the benefits of supporting security through diversification of exporters. LNG has played a positive role in the UK compared with other Eastern and Central European countries that had their gas supply cut off by Russia. Nonetheless, the gas industry missed an opportunity within the public discourse to highlight the positive role LNG has played in energy security.

In National Grid's 2019 "future energy steady progression" and 2°C scenarios, where natural gas use increases without the production of domestic shale gas, LNG imports steadily increase from current levels. This relates directly to a forecasted increase in import dependency up to 2050 (National Grid, 2019a). In other words, for future import dependency to reduce or remain similar to current levels, consumer evolution and community renewable pathways would have to develop, which would imply an unlikely combination of a significant increase in domestic shale production, scaling up of CCS for gas CH_4 reforming to produce hydrogen, and increased use of renewable energy for electrification and electrolysis-produced hydrogen. Thus, the UK's import dependency is likely to increase in the coming decades (National Grid, 2023). That said, the predicted increase in global LNG supply by 2040 (IGU, 2021) may alter the previously perceived risks of increased exposure to LNG.

Overall, LNG will continue to play an important role for energy security reasons because it provides flexibility in an energy system that caters for increasing renewable-energy-produced electricity, varying pathways for gas and energy demand, declining UK and European natural gas production, declining domestic storage capability, and seasonal variations. Exposure to the LNG market, however, also increases exposure to fluctuating global prices, which could affect security of supply. The exact picture remains unclear because the pace of decarbonisation for the rest of Europe will affect overall demand for natural gas and the general attractiveness of the European market for exporters. This, in turn, could affect the competitiveness of natural gas in the UK.

Competitiveness

In the 1950s and 1960s, the state ownership model supported the rapid expansion of electricity capacity. The late 1970s ushered in a transformation from a coal-dominated, state-centric, export-driven energy system to one that was market led, and eventually gas dependent and import focused (Li & Pye, 2018). The UK market developed further in the 1980s and 1990s, against the backdrop of a neoliberal tide and an assertive European Commission, which pushed for the restructuring of the European energy industry by accelerating liberalisation and privatisation and increasing open competition across markets. In the early 1990s, the Government took a hands-off regulatory approach, disbanding the Department of Energy in 1992, essentially downgrading energy policy to a subdivision of the Department of Trade and Industry (Geels et al., 2016). This pro-market policy model continued into the first decade of the 2000s with energy viewed as a tradable commodity and the market as the appropriate vehicle for managing supply. The government's primary role was to establish and enforce fair market rules, with companies determining the energy mix (Kern et al., 2014). The energy policy objectives at this time were to increase market competition and encourage free-market principles to achieve energy affordability. Prompted by market liberalisation, availability of resources, and improved turbine technology, the "Dash for Gas" took place in the early 1990s. In parallel, anti-pollution legislation increased the number of companies competing with coal by allowing regional electricity utility providers to purchase gas-fired electricity and pass on the cost to customers, indicating the policy was about diversifying the market and reducing prices (Foxon, 2013; Trutnevyte, 2016).

Pro-market policies encouraged investment in gas, which led to gas-on-gas competition underpinned by a highly liquid market (Tusiani & Shearer, 2016). In 2017, the UK traded 1,853 BCM of gas at a churn rate of 23, with 142 licensed entities dealing in the National Balancing Point wholesale gas market in 2016 (IEA, 2019). That year, the country had the fourth-lowest gas price for industry and the eighth lowest for households among IEA countries, with a comparatively low 3–5% share of tax in the natural gas price. In the same year, gas bills for end-users comprised wholesale costs (30%), network costs (25%), operating costs (20%), and a pre-tax supply margin (9%) (IEA, 2019).

The economic impact on UK consumers and businesses is a key challenge to achieving net-zero. From this perspective, the natural gas industry could play a positive role if it positioned itself as a competitive element of the energy supply. Market players may have missed an opportunity to focus on the increasing competitiveness of gas as a major selling point. Policymakers need to be more creative and approach the green agenda in a way that does not harm consumers, with end-users shouldering the additional costs of switching to lower-carbon heating without clear direct benefits. This became a greater challenge in the economic slowdown following Brexit and COVID-19, along with doubt about the regulator's ability to successfully manage gas policy in the context of decarbonisation versus market promotion, and manage the financing of efforts to convert the grid and invest in new infrastructure and technology. The UK Government was reactive to public and media sentiment, and messaging inconsistencies can lead to challenges in implementing potentially difficult policies in the future related to cost. For instance, the CCC (2019, p. 28) has been able to make a strong case that mitigation investment is better than adaptation from an economic perspective, with a view towards 2050, and estimated that the cost to achieve emissions reductions of 100% would be between 1% and 2% of GDP, taking into consideration that the actual impact on UK GDP could be lower or even positive. Public perception and awareness may not be at the level required for more far-reaching decisions. However, the Government would struggle to make the case of spending money now. Measures seen as increasing the overall cost of energy for consumers favour natural gas in the short term because consumers have grown accustomed to using and paying for natural gas. UK Census data (House of Commons Library, 2023) indicate that approximately 74% of households use mains gas central heating, 9% rely on electric central heating, 3% use oil central heating, and 1% have no central heating. Currently, new-build developers in the country have a choice of heating systems ranging from modern gas boilers, electric heating, and biomass boilers. However, from 2025, new UK building regulations stipulate that heating systems that burn fossil fuels (i.e. oil, LPG, and natural gas boilers) will be banned from new builds (HM Government, 2024). Thus, new-build properties in the future will probably have heat pumps in place of fuel-burning boilers.

The UK had all the ingredients for a thriving LNG import industry from 2000 to 2015 – increased natural gas demand for power, declining domestic production, security of supply policy drivers, and the liberalisation of its gas market, supported by competitive prices for consumers (Corbeau & Ledesma, 2016). With increased import capacity and the attractiveness of an easy-to-access market, the UK positioned itself to attract LNG suppliers to respond to price signals, which broadly supported security of supply concerns and ensured alignment with global prices. There are only two indices where price is not primarily driven by domestic supply-and-demand concerns – the UK's National Balance Point (NBP) and the Dutch Title Transfer Facility (TTF) – which covers many north-west European countries. US- and Asia-based indices, particularly those in Japan and Korea, respond directly to local dynamics. The NBP and the TTF have evolved to become the global balancing markets, partly owing to European storage and demand conditions but mostly because of global dynamics. The decarbonisation policy approaches and pathways will affect the economics of natural gas alternatives and future competitiveness of natural gas in the UK.

Energy-transition dynamics

Multi-level perspective analysis

Landscape analysis

Several global and European trends have affected the UK's energy landscape in recent decades. The Fourth Industrial Revolution or digital revolution is currently taking place and forms part of a broader global transition supported by disruptive technology such as robotics and artificial intelligence (AI) (CCC, 2019). This revolution also coincides with a change towards a more sustainable way of life across the UK and Europe, supported by political decisions across the EU focusing on decarbonisation and an overall reduction in GHG emissions.

Public awareness of environmental issues and interest in climate change have increased pressure on UK policymakers to develop and implement more progressive environmental policies. In 1990, in a speech to the Royal Society, Prime Minister Margaret Thatcher spoke about climate change as a global challenge and mentioned using economic tools to address pollution (Pearson & Watson, 2012; Trutnevyte, 2016). Since then, the UK's market-driven energy policy has faced a gradually increasing landscape pressure, driven by government commitments to national and international emissions targets and concerns around affordability, economic development, and security of supply (Foxon, 2013).

In a series of public speeches between 2003 and 2004, the Government's Chief Scientific Adviser, David King, claimed that the threat posed by climate change to the world is more serious than terrorism (Carter, 2014). In 2005 and 2006, climate change gained greater political and public attention following the release of the Stern Review on the economics of climate change and Al Gore's popular film "An Inconvenient Truth" (Carter & Jacobs, 2014; Stern, 2006). Climate change gradually gained prominence over the next decade and, according to the BEIS attitudes tracker, 82% of the public was fairly concerned about climate change. This figure has grown continuously, with nearly 70% of those surveyed who believe in climate change saying that they can feel its impact in the UK, highlighting that this is a present and active issue (BEIS, 2019c, 2023). The emergence of a youth-led movement focused on environmental issues has sparked the emergence of other movements, such as Extinction Rebellion, which have had some success in raising climate-related issues within the public discourse. Environmental groups have argued that the initial carbon budgets were not sufficiently aggressive and that some options such as nuclear energy and CCS are not desirable long-term options.

It remains to be seen how the factors that led to Brexit and the overall rise of populism throughout Europe and the UK could impact potential energy-transition pathways. Social and political issues linked to the rise of nationalism and new barriers to global trade could also prove to be barriers to transition (Whinskel et al., 2019), as witnessed, for example, by the Trump administration's withdrawal from the Paris Agreement in 2017. Current and future governments will have to balance public sentiment and economic considerations, which could have an impact on climate-related policy goals.

Innovation and R&D ecosystem

Despite the UK's long history in energy-related R&D, overall spending has steadily decreased in recent decades, with R&D spending representing 1.69% of GDP in 2017, well below the OECD average. At the same time, energy-related government spending stood at £536 million – approximately 0.26% of GDP and below the IEA's average of 0.35% (IEA, 2019). The UK Government, however, plans to increase its R&D spending to 2.4% of GDP by 2027, with increased focus on decarbonisation (IEA, 2019). The energy R&D framework operates under the Energy Innovation Board, which offers a platform from which to coordinate and fund policies and programmes across government agencies. In 2018, a merger between Innovate UK and the research councils created UK Research and Innovation, a public entity sponsored by BEIS with a mandate to focus on science and academic collaboration.

The UK's energy innovation and research policies focus on the development of clean-growth industries, reflected in the UK Industrial Strategy, aid strategy, and Clean Growth Strategy, the last of which was launched in 2017 to promote energy technology development and innovation to decarbonise the energy system. Clean growth and the future of mobility were two of the four grand challenges in the UK's Industrial Strategy, alongside AI and data and an ageing society (BEIS, 2017). The aims of the clean-growth challenge include halving the energy consumption of new buildings by 2030 and establishing the world's first net-zero carbon industrial cluster by 2040, with at least one low-carbon cluster by 2030. The UK has made efforts to align the country's R&D agenda to the UK's net-zero objective. However, investments do not match the scale and ambition required, with interviewees pointing to other countries such as Japan which are consistently ranked in the OECD's top three for energy R&D spending. The UK's progress has been slowed by periods of austerity. In the future, competing priorities following Brexit and adaptation measures required by climate change will put additional pressure on budgets.

In the early 2010s, new technology-specific innovation systems were supported by niche-level government policies, including at-scale onshore and offshore wind, and small-scale solar photovoltaics (Foxon, 2013). While the focus on renewables is forecast to continue, hydrogen and CCS are two promising areas that could help the UK meet its domestic need and provide a platform for companies and researchers to have a global impact. The UK's R&D plans can be linked to international developments, with the country's direction dictated to by global trends, as the Government look at potential markets and markets in development. Moreover, while it may be difficult to make a sizeable global impact based on the scale of the research, the UK could excel in both hydrogen and CCS technology development by building on its existing skills and infrastructure. The country could become a key global player with the support of appropriate policies that leverage the UK's strengths over the long term. Although at this stage hydrogen technologies in the UK have a long way to go before enabling a rapid transition, the sector meets several of Geels and Schot's (2007) criteria for reaching innovation maturity, including the presence of powerful actors, improving economics, and increasing performance.

Regime status summary

The UK's energy system has responded positively to energy efficiency measures and decarbonisation efforts through an increase in fuel-switching and the use of renewables in recent decades. Nonetheless, the main components of the old regime persist, namely, heavily centralised and carbon-intensive energy production. Well-established infrastructure results in a system that is still dominated by natural gas for space heating and, to a lesser extent, electricity. Climate activists and many academics in the UK consider that more fundamental changes are needed than have occurred in the past 50 years, including a total transformation and reimagination of how energy is used, transported, and produced (Li & Pye, 2018). Alongside full electrification and decentralisation, other paths towards net-zero have been developed, where decarbonisation, with the support of natural gas and CCS, can facilitate the transition by leveraging existing infrastructure and implementing all the above policies. Interviewees cited in Al-Kuwari (2023) noted that offshore wind farms and, in the future, industrial clusters bring into question the notion that the energy transition will probably lead to a decentralised energy system and that the economics and impracticalities of full electrification position decarbonisation as part of a future low-carbon energy regime.

MLP analysis for the UK identifies several significant landscape pressures that have influenced climate discourse and energy policy. CCS and hydrogen have the potential to stimulate technology niches in the medium to long term, with increased funding, targeted sectoral policies, and appropriate R&D and innovation support. However, the success of CCS and hydrogen in the near term is to some extent dependent on the continued use of natural gas and the ability to leverage natural gas infrastructure, providing LNG with a window of opportunity. Increased electrification through renewables and nuclear energy could disrupt the future viability of natural gas in the UK; however, limitations on storage and the ability to address hard-to-abate non-power sectors limit the potential for significant regime disruptions in the UK in the coming decade without major improvements in green hydrogen technology.

Transition context

This section analyses the readiness and future energy-transition trajectory of the UK's existing system using an adapted version of Smith et al.'s (2005) governance of sustainability transition framework. The two main criteria for evaluation are the UK's energy-transition ambition and its adaptive capacity.

Energy-transition ambition

The UK is a global leader in setting decarbonisation targets. The 2008 Climate Change Act established legally binding carbon budgets that set limits on net amounts of GHG emissions over successive five-year terms, to be monitored by energy and emissions projections (BEIS, 2019d). The Foreword to the UK's 2009 Low Carbon Transition plan stated that the transition to a low-carbon economy will be a defining issue of the twenty-first century (HM Government, 2009). In 2019, the UK committed to net-zero emissions by 2050, the first major economy in the world to commit to legally binding targets to end its contribution to global warming.

Adaptive capacity

The UK shows a high degree of adaptive efficiency as an open-access social order amenable to forms of competition that provide opportunities to develop new solutions to new problems (Andrews-Speed, 2016). The evolution of the energy market and introduction of technologies and applications, supported by innovative policies over the decades, have demonstrated this. The UK's core strengths to support such a transition are the North Sea infrastructure, with its storage potential and associated developed workforce and know-how. Those within the energy industry are key components of that strength, with offshore expertise providing a strong base from which to focus on technologies such as CCUS and hydrogen to leverage current onshore and offshore infrastructure. Several of Al-Kuwari's (2023) interviewees identified that major UK oil and gas operators, and others across the supply chain, have local knowledge and expertise that can be applied to the low-carbon transition. CCS could be an area of leadership for these companies. In the UK, CCS may be achieved most successfully by one of the oil and gas companies because these operators could apply their skills to remove and store large amounts of CO_2. Moreover, the energy industry can continue to develop large-scale wind energy projects. Although these are often viewed as unpopular and expensive, they have been successful because there are a "relatively small number of decisions and a small number of players making significant investments" (Al-Kuwari, 2023). This quick ramp-up and scale could be replicated elsewhere by leveraging this expertise and learning from the policies that have contributed to rapid cost reductions (Grubb & Newbery, 2018).

Policymakers in the UK have demonstrated an ability to innovate and adapt energy- and environmental-related policies, despite a deep-rooted institutional inertia in a political system that has largely been in place since the thirteenth century. In the UK parliamentary system, matters which have an economic interest are generally managed by closed producer groups. While a network of industry members and policymakers allows for open interaction, new entrants and external stakeholders have difficulty accessing it (Carter, 2014). A political structure that maximises the likelihood of a strong executive government can be seen as advantageous insofar as a single-party parliamentary majority can effect significant change. In contrast, coalition governments can create opportunities for cross-party consensus, as seen during the Conservative–Liberal Democrat coalition government that approved the fourth carbon budget in 2011 during the era of "consensus politics", when the three main political parties supported progressive climate policies with strong public and media support (Carter, 2014; Geels et al., 2016).

Two centuries of energy and industrial policy evolved into distinct phases in the UK, with transitions triggered by financial constraints, war, or structural inefficiencies. The phases included the competitive era, the regulation and municipalisation era, the rationalisation and nationalisation eras, and finally the privatisation and liberalisation era of the 1990s (Foreman-Peck & Millward, 1994). The policy paradigm that emerged in 2010 derived from previous market-based principles with recognition that the Government had to adopt a more active, strategic role. Energy security and the need to meet legally binding climate targets became main priorities, overtaking affordability (Kern et al., 2014). In the recent era of climate change and decarbonisation, what began as a series of high-level goals soon became specific targets, demonstrating the Government's ability to implement innovative and effective policies. In 2009, the UK Low Carbon Transition plan for the electricity sector set out a domestic target of 30% electricity generation from renewable sources by 2020, with near full decarbonisation by 2030. Between 2009 and 2012, the amended Renewables Obligation, UK Renewable Strategy, Carbon Plan, and Energy Bill were announced to promote policy delivery (Geels et al., 2016). The Electricity Market Reform initiative in 2013 addressed issues around delivery of an increasing share of energy from renewable sources by combining competitive auctions for firm capacity and renewable energy and introducing a set carbon price to stimulate technological change and restructure the sector (Grubb & Newbery, 2018). At a European level, the UK was credited with being a key promoter of market-based competition innovation – ideas that led to the EU Gas and Electricity Directives in the 1990s and first decade of the 2000s (Gustafson, 2020). The neoliberal approach was based on open access and improved competition, although it had a negative impact for domestic electricity market incumbents' financial standing.

The UK's strong legal and regulatory infrastructure has allowed it to transition out of the EU while maintaining momentum on energy-transition-related policies. Some have questioned the political will and long-term ability for current and future governments to deliver on bold objectives. Perceived setbacks related to Brexit and the unsuccessful initial management of the COVID-19 pandemic contributed to a weakened macroeconomic status, limiting the Government's ability to deliver the necessary energy transition in a politically acceptable manner. While some of Al-Kuwari's (2023) interviewees pointed to missteps related to the transport sector, CCS, and nuclear tariff discussions, there was also acknowledgement of the UK's ability to implement effective measures with demonstrably strong results. Progressive policies related to a reduction in overall costs of large offshore wind projects and an ability to rapidly reduce coal's presence in the electricity market were noted as successes. One overarching feature of these policies is the willingness to test new energy propositions, which could provide the UK with a distinct advantage over its European counterparts.

The City of London is a global financial hub with the strength to facilitate and drive investment in new initiatives and large-scale projects. The City is a key participant in the development of the green economy, and despite Brexit, remains a strong, vibrant finance centre that could help facilitate investment in green projects and support government policies on environmental, social, and corporate initiatives. The UK's "all of the above" energy and climate approach has freed policymakers from general technological constraints – a strength in the country's low-carbon transition. Willingness to consider several technologies and solutions, underpinned by long-term goals, is exemplified by the "net-zero, not zero" and the "molecules vs electrons" non-debate. The system includes top-down and centralised components owing to institutional inertia. However, it is also progressive and open to taking risks around interactions between industry, government, and financial institutions, all influenced by an active media and engaged civil society. In the complex, fast-changing socio-economic environment, the UK's policymaking ability to adapt will play a key role in supporting the country's energy-transition ambition.

Geels et al. (2016) concluded that the UK is on a "transformation pathway" with incumbent actors deploying large-scale renewable energy. The UK's political system, liberal market economy, and a stagnating industrial and manufacturing sector encourage a pathway in which incumbents will continue to play a role. Landscape pressures stimulate a gradual regime adjustment, where niche innovations are not yet sufficiently developed to make a significant impact (Geels et al., 2016). Based on Smith et al.'s (2005) transition context methodology, and considering its adaptive capacity levers and selection pressures, the UK finds itself in a purposive transition with certain elements pushing it towards "endogenous renewal" such as the ability to leverage North Sea infrastructure and resources. Strong landscape pressures have been identified, alongside some niche innovations that have the potential to disrupt the current system in the future. Organisations in this sector may also reorient towards radical niche innovations, allowing them to incrementally adjust. The broad social, political, and business support for the net-zero agenda, coupled with a strong regulatory framework, lead the UK towards a purposive transition. "Leading by example" supports the national approach towards a low-carbon transition, although there are still debates to be held regarding the pathways and trade-offs in decarbonisation and climate policies.

Actors

Several international elements and agreements influence the UK energy system, including the multinational ownership structure of energy companies, non-governmental organisations (NGOs), the proportion of imported energy, and EU and global climate commitments. The market reforms of the 1990s led to privatisation and liberalisation of large parts of the UK energy system, shifting the balance of power, albeit with a significant element of centralisation remaining (Bridge et al., 2013; EIA, 2018; HM Government, 2023). The key participants influencing the development of the UK natural gas sector are the Government and government-related entities, major integrated energy companies, influencers who have direct and indirect impact on the UK gas market (such as NGOs), media and academics, and those who export large amounts of gas via LNG or pipeline to the UK.

Summary of the T-E-T-I analysis

Figure 3.1 provides a visual summary of the UK case study within the T-E-T-I framework. The UK case highlights the factors that determine LNG's potential future pathways within the country's energy system in the context of the current energy transition. For policymakers and large utilities, among others, LNG provides a flexible solution for switching away from coal while leveraging well-developed

Resources

- Stagnating domestic gas reserves; increasing renewables
- Coal no longer a factor; uncertain nuclear future pathway
- Established pipeline and LNG infrastructure, however, lacks natural gas storage
- Key building blocks in place for potential CCS and hydrogen, including capabilities
- LNG plays important seasonal role to support renewables, key for heating sector

Market

- Supports market attractiveness
- Funding for fossil fuel development decreasing rapidly
- Trade discussion and impact of Brexit could impact the UK gas market's role and attractiveness
- Gas relatively affordable vs OECD and EU; renewables' role rapidly growing in the power sector
- Seasonality role for LNG may limit further investments into sector

Energy Transition Dynamics

- Social movement – climate change as 'moral issue,' policymakers have public support on net-zero agenda
- Potential for CCS and hydrogen could eventually lead to disruptive niche
- Lack of deep funding and no path to increase R&D and innovation in short to medium term
- 'Low hanging fruit' stage completed, renewable ramp-up significant; hard-to-abate sectors, large-scale storage will be more challenging

Power

- Strong regulatory system in place
- Politicians deal with competing pressures, drivers of energy policy shifting
- Stagnating domestic gas reserves; increasing renewables
- Challenge to deliver major infrastructural changes owing to Brexit, political gridlock, NIMBYism and lack of funds
- 'Big 6' power dwindling, less impact on energy policy direction
- Gas lobby weakened by 'green' narrative

Figure 3.1 Summary of the UK case study within the T-E-T-I framework (developed by Al-Kuwari, 2020)

gas-related infrastructure. In the medium term, LNG can continue to meet gas demand and support the development of renewables, thus maintaining the country's energy security while adapting to seasonal demand fluctuations. Longer term, the role of LNG is unclear and will, to some extent, be determined by regional market developments, technological improvements in hydrogen production and CCS, the role of nuclear, and global climate action. There is also potential for the LNG industry to adopt a more proactive approach to leverage existing and potential future infrastructure and build on the public's acceptance of natural gas as a fuel source. Given environmental and operational improvements to decarbonise its own value chain, the LNG industry can play an important role in the development of hydrogen and, thus, take on a key role in the UK's decarbonisation transition (Al-Kuwari, 2022).

The UK has been notably successful in phasing out coal and developing a home-grown gas market to support its energy security goals, make gas more affordable, and reduce overall GHG emissions. The UK was able to work in conjunction with the EU's gas and power markets and regulations to meet its goals. In addition, an overall period of economic growth and relative political stability supported its abilities to introduce and implement sound policies. Going forward, UK policymakers face a more volatile environment with new challenges, while having to meet challenging "trilemma" dynamics. Post-COVID-19 and Brexit economic challenges required new infrastructure, including retrofitting, additional nuclear, and a volatile political environment – all pressures that are likely to make radical changes possible. LNG may prove to be a stable tool to deliver much-needed energy, subject to affordability concerns in the short term, and its environmental footprint in the medium to longer term.

Note

1 BEIS existed until 2023 when it was split to form the Department for Business and Trade (DBT), the Department for Energy Security and Net Zero (DESNZ), and the Department for Science, Innovation and Technology (DSIT). Responsibility for national security and investment policy has gone to the Cabinet Office.

References

Al-Kuwari, O. (2020, May 18). *Our new energy transformation can see the Middle East thrive – if it invests in RDI and education.* Qatar Foundation Op-Ed. www.qf.org.qa/stories/op-ed-our-new-energy-transformation-can-see-the-middle-east-thrive-if-it

Al-Kuwari, O. (2022). LNG's role in a decarbonising energy system: Lessons from the United Kingdom. In D. S. Olawuyi & E. G. Pereira (Eds.), *The Palgrave handbook of natural gas and global energy transitions* (pp. 141–171). Palgrave Macmillan.

Al-Kuwari, O. (2023). *The future of liquified natural gas (LNG) in the energy transition: Options and implications for the LNG industry in a decarbonising world* [Unpublished doctoral dissertation]. University College London. https://discovery.ucl.ac.uk/id/eprint/10173650/

Andrews-Speed, P. (2016). Applying institutional theory to the low-carbon energy transition. *Energy Research and Social Science*, *13*, 216–225. https://doi.org/10.1016/j.erss.2015.12.011

Arapostathis, S., Carlsson-Hyslop, A., Pearson, P., Thornton, J., Gradillas, M., Laczay, S., & Wallis, S. (2013). Governing transitions: Cases and insights from two periods in the history of the UK gas industry. *Energy Policy*, *52*, 25–44. https://doi.org/10.1016/j.enpol.2012.08.016

BEIS (Department of Business Energy & Industrial Strategy). (2017). *The clean growth strategy – leading the way to a low carbon future.* HM Government. https://assets.publishing.service.gov.uk/media/5ad5f11ded915d32a3a70c03/clean-growth-strategy-correction-april-2018.pdf

BEIS (Department of Business Energy & Industrial Strategy). (2018). *Clean growth – transforming heating.* HM Government. https://assets.publishing.service.gov.uk/media/5c191c05ed915d0b9211b9c5/decarbonising-heating.pdf

BEIS (Department of Business Energy & Industrial Strategy). (2019a). *BEIS public attitudes tracker review (wave 29).* HM Government. https://assets.publishing.service.gov.uk/government/uploads/system/uploads/attachment_data/file/800429/BEIS_Public_Attitudes_Tracker_-_Wave_29_-_key_findings.pdf

BEIS (Department of Business Energy & Industrial Strategy). (2019b). *BEIS public attitudes rracker review (wave 30).* HM Government. https://assets.publishing.service.gov.uk/media/5d2efdce40f0b64a8099e1c3/BEIS_Public_Attitudes_Tracker_Wave_30_cross_tabulation_tables_Excel.xlsx

BEIS (Department of Business Energy & Industrial Strategy). (2019c). *BEIS public attitudes tracker review (wave 31).* HM Government. www.gov.uk/government/collections/public-attitudes-tracking-survey#beis-public-attitudes-tracker-surveys

BEIS (Department of Business Energy & Industrial Strategy). (2019d). *Updated energy and emissions projections for the UK 2018. The energy white paper.* HM Government. www.gov.uk/government/publications/updated-energy-and-emissions-projections-2018

BEIS (Department of Business Energy & Industrial Strategy). (2023). *BEIS public attitudes tracker review (winter 2022).* HM Government. https://assets.publishing.service.

gov.uk/media/640704618fa8f527f2ef965b/BEIS_PAT_Winter_2022_Energy_Infrastructure_and_Energy_Sources.pdf

BP (British Petroleum). (2019a). *BP energy outlook 2019*. www.bp.com/content/dam/bp/business-sites/en/global/corporate/pdfs/energy-economics/energy-outlook/bp-energy-outlook-2019.pdf

BP (British Petroleum). (2019b). *BP statistical review – 2019: Global energy market in 2018*. www.bp.com/content/dam/bp/business-sites/en/global/corporate/pdfs/energy-economics/statistical-review/bp-stats-review-2019-full-report.pdf

Bridge, G., Bouzarovski, S., Bradshaw, M., & Eyre, N. (2013). Geographies of energy transition: Space, place and the low-carbon economy. *Energy Policy*, *53*, 331–340. https://doi.org/10.1016/j.enpol.2012.10.066

Bridge, G., & Bradshaw, M. (2017). Making a global gas market: Territoriality and production networks in liquefied natural gas. *Economic Geography*, *93*(3), 215–240. https://doi.org/10.1080/00130095.2017.1283212

CarbonBrief. (2023, June 28). *CCC: Chance of UK meeting climate pledges has 'worsened' since last year*. www.carbonbrief.org/ccc-chance-of-uk-meeting-climate-pledges-has-worsened-since-last-year/

Carter, N. (2014). The politics of climate change in the UK. *Wiley Interdisciplinary Reviews: Climate Change*, *5*(3), 423–433. https://doi.org/10.1002/wcc.274

Carter, N., & Jacobs, M. (2014). Explaining radical policy change: The case of climate change and energy policy under the British labour government 2006–10. *Public Administration*, *92*(1), 125–141. https://doi.org/10.1111/padm.12046

CCC (Committee on Climate Change). (2016). *The future of carbon capture and storage in the UK. House of commons, second report*. House of Commons. https://publications.parliament.uk/pa/cm201516/cmselect/cmenergy/692/692.pdf

CCC (Committee on Climate Change). (2018). *Hydrogen in a low-carbon economy*. www.theccc.org.uk/wp-content/uploads/2018/11/Hydrogen-in-a-low-carbon-economy.pdf

CCC (Committee on Climate Change). (2019). *Net zero: The UK's contribution to stopping global warming*. www.theccc.org.uk/publication/net-zero-the-uks-contribution-to-stopping-global-warming/

Corbeau, A., & Ledesma, D. (Eds.). (2016). *LNG markets in transition: The great reconfiguration*. Oxford University Press for the Oxford Institute for Energy Studies and the King Abdullah Petroleum Studies and Research Center.

Department for Energy Security and Net Zero. (2023a, July 27). *Digest of UK energy statistics (DUKES) 2023. Chapter 4: Natural gas*. https://assets.publishing.service.gov.uk/media/64f1fc589ee0f2000db7bdd7/DUKES_2023_Chapter_4_Gas.pdf

Department for Energy Security and Net Zero. (2023b, March 30). *Supply of liquefied natural gas in the UK, 2022*. https://assets.publishing.service.gov.uk/media/642434bd3d885d000cdade0f/Supply_of_Liquefied_Natural_Gas_in_the_UK__2022.pdf

Department for Energy Security and Net Zero. (2024a). *2022 UK greenhouse gas emissions, final figures*. https://assets.publishing.service.gov.uk/media/65c0d15863a23d0013c821e9/2022-final-greenhouse-gas-emissions-statistical-release.pdf

Department for Energy Security and Net Zero. (2024b). *Final statement for the third carbon budget. Reporting UK 2018–2022 emissions to parliament under the climate change act 2008*. https://assets.publishing.service.gov.uk/media/664356ef4f29e1d07fadc6e9/third-carbon-budget-statement.pdf

Department of Trade and Industry. (2003). *Energy white paper – our energy future: Creating a low carbon economy*. https://assets.publishing.service.gov.uk/media/5a7c1f5940f0b645ba3c6d4f/5761.pdf

Dodds, P. E., & Demoullin, S. (2013). Conversion of the UK gas system to transport hydrogen. *International Journal of Hydrogen Energy*, *38*(18), 7189–7200. https://doi.org/10.1016/j.ijhydene.2013.03.070

EIA (Energy Information Administration). (2018). *Country analysis brief – United Kingdom*. www.eia.gov/international/analysis/country/GBRU.S.

EIA (Energy Information Administration). (2024). *Country analysis brief – United Kingdom.* www.eia.gov/international/content/analysis/countries_long/United_Kingdom/pdf/UK%20CAB_2024.pdf

Foreman-Peck, J., & Millward, R. (1994). *Public and private ownership of British industry 1820–1992*. Clarendon Press. https://doi.org/10.1093/acprof

Fouquet, R. (2010). The slow search for solutions: Lessons from historical energy transitions by sector and service. *Energy Policy*, *38*(11), 6586–6596. https://doi.org/10.1016/j.enpol.2010.06.029

Foxon, T. J. (2013). Transition pathways for a UK low carbon electricity future. *Energy Policy*, *52*, 10–24. https://doi.org/10.1016/j.enpol.2012.04.001

Geels, F. W. (2011). The multi-level perspective on sustainability transitions: Responses to seven criticisms. *Environmental Innovation and Societal Transitions*, *1*(1), 24–40. https://doi.org/10.1016/j.eist.2011.02.002

Geels, F. W., Kern, F., Fuchs, G., Hinderer, N., Kungl, G., Mylan, J., Neukirch, M., & Wassermann, S. (2016). The enactment of socio-technical transition pathways: A reformulated typology and a comparative multi-level analysis of the German and UK low-carbon electricity transitions (1990–2014). *Research Policy*, *45*(4), 896–913. https://doi.org/10.1016/j.respol.2016.01.015

Geels, F. W., & Schot, J. (2007). Typology of sociotechnical transition pathways. *Research Policy*, *36*(3), 399–417. https://doi.org/10.1016/j.respol.2007.01.003

German Federal Government. (2024). *The national hydrogen strategy*. www.bmwk.de/Redaktion/EN/Hydrogen/Dossiers/national-hydrogen-strategy.html

Gouiffes, P. (2007). *Margaret Thatcher & the miners*. www.scribd.com/document/94955157/Margaret-Thatcher-and-the-Miners

Grubb, M., & Newbery, D. (2018). *UK electricity market reform and the energy transition: Emerging lessons*. MIT Center for Energy and Environmental Policy Research. https://ceepr.mit.edu/wp-content/uploads/2021/09/2018-004.pdf

Gustafson, T. (2020). *The bridge: Natural gas in a redivided Europe*. Harvard University Press.

Hinson, S., Sutherland, N., Rhodes, C., Bolton, P., & Cracknell, R. (2020). *UK oil and gas industry debate (pack CDP 2020/0036)*. House of Common Library. https://researchbriefings.files.parliament.uk/documents/CDP-2020-0036/CDP-2020-0036.pdf

HM Government. (2009). *The UK low carbon transition plan*. https://assets.publishing.service.gov.uk/media/5a74b4b1e5274a3cb2866852/9780108508394.pdf

HM Government. (2021). *UK hydrogen strategy*. https://assets.publishing.service.gov.uk/media/64c7e8bad8b1a70011b05e38/UK-Hydrogen-Strategy_web.pdf

HM Government. (2022, April 7). *Policy paper: British energy security strategy*. www.gov.uk/government/publications/british-energy-security-strategy/british-energy-security-strategy#energy-plan-objectives-and-key-measures

HM Government. (2023). *Powering up Britain. Energy security plan*. https://assets.publishing.service.gov.uk/media/642708eafbe620000f17daa2/powering-up-britain-energy-security-plan.pdf

HM Government. (2024, March 4). *The future homes and buildings standards: 2023 Consultation*. www.gov.uk/government/consultations/the-future-homes-and-buildings-standards-2023-consultation/the-future-homes-and-buildings-standards-2023-consultation

House of Commons Library. (2023, June 19). *Constituency data: Central heating, 2021 census*. https://commonslibrary.parliament.uk/constituency-data-central-heating-2021-census/#:~:text=Most%20households%20in%20England%20and,1%25%20had%20no%20central%20heating

IEA (International Energy Agency). (2017). *World energy investment 2018*. IEA Publications. https://iea.blob.core.windows.net/assets/3688404e-dc52-46be-b347-d2d211799818/WEI2018.pdf

IEA (International Energy Agency). (2018). *World energy outlook 2018*. IEA Publications. www.iea.org/reports/world-energy-outlook-2018

IEA (International Energy Agency). (2019). *Energy policies of IEA countries: United Kingdom 2019 review*. https://iea.blob.core.windows.net/assets/298930c2-4e7c-436e-9ad0-2fb8f1cce2c6/Energy_Policies_of_IEA_Countries_United_Kingdom_2019_Review.pdf

IEA (International Energy Agency). (2023, December 21). *United Kingdom*. www.iea.org/countries/united-kingdom/energy-mix

IGU (International Gas Union). (2021). *World LNG report 2021*. www.igu.org/resources/world-lng-report-2021/

IMF (International Monetary Fund). (2024, April). *World economic outlook database*. www.imf.org/en/Publications/WEO/weo-database/2024/April

Kern, F., Kuzemko, C., & Mitchell, C. (2014). Measuring and explaining policy paradigm change: The case of UK energy policy. *Policy and Politics*, *42*(4), 513–530. https://doi.org/10.1332/030557312X655765

Krug, M., & Lebelhuber, C. (2018). *Study on the future role of gas from a regulatory perspective*. Council of European Energy Regulators. www.ceer.eu/wp-content/uploads/2024/05/CEER-Report-of-Future-Role-of-Gas.pdf

Li, F. G. N., & Pye, S. (2018). Uncertainty, politics, and technology: Expert perceptions on energy transitions in the United Kingdom. *Energy Research and Social Science*, *37*, 122–132. https://doi.org/10.1016/j.erss.2017.10.003

McDowall, W. (2014). Exploring possible transition pathways for hydrogen energy: A hybrid approach using socio-technical scenarios and energy system modelling. *Futures*, *63*, 1–14. https://doi.org/10.1016/j.futures.2014.07.004

McGlade, C., Bradshaw, M., Anandarajah, G., Watson, J., & Ekins, P. (2014). *A bridge to a low-carbon future? Modelling the long-term global potential of natural gas*. UK Energy Research Centre. www.ucl.ac.uk/bartlett/sustainable/sites/bartlett_sustainable/files/gas-as-a-bridge.pdf

McGlade, C., Pye, S., Ekins, P., Bradshaw, M., & Watson, J. (2018). The future role of natural gas in the UK: A bridge to nowhere? *Energy Policy*, *113*, 454–465. https://doi.org/10.1016/j.enpol.2017.11.022

National Grid. (2019a, July). *Future energy scenarios*. www.nationalgrideso.com/document/170756/download

National Grid. (2019b). *Gas ten year statement 2019*. www.nationalgas.com/document/128886/download

National Grid. (2022). *Future energy scenarios 2022*. www.nationalgrideso.com/future-energy/future-energy-scenarios-fes

National Grid. (2023). *Future energy scenarios 2023*. www.nationalgrideso.com/document/283101/download

Offenberg, P. (2016). *The European neighbourhood and the EU's security of supply with natural gas*. Jacques Delors Institute. https://institutdelors.eu/en/publications/the-european-neighbourhood-and-the-eus-security-of-supply-with-natural-gas/

ONS (Office for National Statistics). (2024, May 10). *Gross domestic product: Year on year growth*. www.ons.gov.uk/economy/grossdomesticproductgdp/timeseries/ihyp/pn2

Pearson, P., & Watson, J. (2012). *UK energy policy 1980–2010. A history and lessons to be learnt. A review to mark 30 years of the parliamentary group for energy studies*. Parliamentary Group for Energy Studies. www.theiet.org/media/9387/uk-energy-policy1980-2010-a-history-and-lessons-to-be-learnt.pdf

Pingkuo, L., & Xue, H. (2022). Comparative analysis on similarities and differences of hydrogen energy development in the world's top 4 largest economies: A novel framework. *International Journal of Hydrogen Energy*, *47*(16), 9485–9503. https://doi.org/10.1016/j.ijhydene.2022.01.038

Pye, S., Li, F. G. N., Price, J., & Fais, B. (2017). Achieving net-zero emissions through the reframing of UK national targets in the post-Paris Agreement era. *Nature Energy*, *2*(3), 1–8. https://doi.org/10.1038/nenergy.2017.24

Pye, S., Li, P. H., Keppo, I., & O'Gallachoir, B. (2019). Technology interdependency in the United Kingdom's low carbon energy transition. *Energy Strategy Reviews*, *24*, 314–330. https://doi.org/10.1016/j.esr.2019.04.002

Sadler, D., Cargill, A., Crowther, M., Rennie, A. J. W., Burton, S., & Haines, M. (2016). *Leeds city gate project H21*. Leeds City Gate. www.northerngasnetworks.co.uk/wp-content/uploads/2017/04/H21-Report-Interactive-PDF-July-2016.compressed.pdf

Selden, Z. (2020). Taming the bear: American liquified natural gas (LNG) exports and the negation of Russian influence in Europe. *Global Affairs*, *6*(2), 149–165. https://doi.org/10.1080/23340460.2020.1780460

Smith, A., Stirling, A., & Berkhout, F. (2005). The governance of sustainable socio-technical transitions. *Research Policy*, *34*(10), 1491–1510. https://doi.org/10.1016/j.respol.2005.07.005

Staffell, I., Scamman, D., Abad, A. V., Balcombe, P., Dodds, P. E., Ekins, P., Shah, N., & Ward, K. R. (2018). The role of hydrogen and fuel cells across the global energy system. *Energy & Environmental Science*, *2*, 463–491. https://doi.org/10.17605/OSF.IO/RZM4G

Stern, J. (2019). *Narratives for natural gas in decarbonising European energy markets OIES paper NG 141*. Oxford Institute for Energy Studies. https://doi.org/10.26889/9781784671280

Stern, N. (2006). *The economics of climate change: The Stern review*. Cambridge University Press.

Toke, D. (2013). Climate change and the nuclear securitisation of UK energy policy. *Environmental Politics*, *22*(4), 553–570. https://doi.org/10.1080/09644016. 2013.806630

Toke, D., & Vezirgiannidou, S. E. (2013). The relationship between climate change and energy security: Key issues and conclusions. *Environmental Politics*, *22*(4), 537–552. https://doi.org/10.1080/09644016.2013.806631

Trutnevyte, E. (2016). Does cost optimization approximate the real-world energy transition? *Energy*, *106*, 182–193. https://doi.org/10.1016/j.energy.2016.03.038

Tusiani, M., & Shearer, G. (2016). *LNG: Fuel for a changing world: A nontechnical guide*. Pennwell Books.

UNFCCC (United Nations Framework Convention on Climate Change). (2015). *Paris Agreement under the United Nations framework convention on climate change*. Retrieved April 22, 2016, from https://unfccc.int/sites/default/files/english_paris_agreement.pdf

Watson, J., Gross, R., Ketsopoulou, I., & Whinskel, M. (2015). The impact of uncertainties on the UK's medium-term climate change targets. *Energy Policy*, *87*, 685–695. https://doi.org/10.1016/j.enpol.2015.02.030

Whinskel, M., Kattirtzi, M., Lowes, R., Woodman, B., & Poulter, H. (2019). *Disrupting the UK energy system: Causes, impacts and policy implications*. UKERC. https://d2e1qxpsswcpgz.cloudfront.net/uploads/2020/03/ukerc_disruption_report_web.pdf

4 Japan

Introduction

Japan has historically been a fuel importer with limited domestic natural resources (Oshiro et al., 2016). It has also been a pioneer in the LNG industry, importing natural gas since the 1970s to improve air quality and provide peak-load power generation to meet industrial demand (Boersma & Jordaan, 2017; Corbeau & Ledesma, 2016). Nuclear power met a large portion of the country's power generation needs from the 1970s as part of a strategy to reduce its dependence on foreign oil following the 1973 OPEC oil crisis (Solomon & Krishna, 2011).

On 11 March 2011, an accident at Fukushima's Daiichi Nuclear Power Plant in Okuma, Fukushima Prefecture was triggered by an earthquake and subsequent tsunami, causing significant regional physical and environmental damage (BfS, 2022, 2024; Efird et al., 2018; Kucharski & Unesaki, 2018). Japanese policymakers had to readjust the country's energy supply system in the immediate aftermath of the accident, which led to a re-emergence of coal in the system and an increase in gas imports to balance supply and demand (Hugo & Kusumawardhana, 2023; Trencher et al., 2020). Subsequent negative public sentiment around nuclear power also forced the Government to reconsider the long-term role of nuclear power in Japan's energy mix. Within this context, LNG provided necessary supplies at short notice, albeit at a high cost, as the volatility exposed the economy to high global prices and affected the Government's future energy policies (Boersma & Jordaan, 2017; Hugo & Kusumawardhana, 2023). Indeed, import dependency for primary energy supply increased to 94% in 2014, its highest level in decades (Hugo & Kusumawardhana, 2023; Oshiro et al., 2016).

These Fukushima accident-related effects triggered the development of a new strategic energy plan (SEP) in 2014, heralding structural and institutional reforms across Japan's energy and climate policies and underpinning a socio-technical transition (Hugo & Kusumawardhana, 2023; Kucharski & Unesaki, 2018). Future energy policy will be shaped by sociopolitical factors, including domestic concerns over safety, the economic impact of fossil fuel imports, global climate change mitigation priorities, and international relations considerations with respect to Japan's desire to maintain its influence and expertise in nuclear power (Len & Nian, 2014).

DOI: 10.4324/9781003545149-4

Although natural gas demand is forecast to increase globally to 2035 and beyond, demand is expected to decrease in Japan (and the EU) (IEA, 2016a, 2018, 2023, n.d.; IEEFA, 2024; McKinsey Global Institute, 2019). However, with nuclear energy's future in doubt, Japan may continue to rely on natural gas as a less-polluting alternative to coal, and LNG is set to play a key role because the country lacks international pipeline gas connections. Longer term, however, LNG's role is unclear in the context of aggressive decarbonisation policies, including a net-zero 2050 pledge and the development of a hydrogen economy. The analysis in this chapter again follows the T-E-T-I framework, discussing resources, market, energy-transition dynamics, and power in turn.

Natural gas and LNG in Japan

History and context

Two energy transitions have taken place in Japan – an economic expansion following the Second World War supported by oil consumption, and a shift to focus on nuclear and gas-fired power generation following the 1970s oil crises, leading to a more diversified energy mix (McLellan et al., 2016). Between 1970 and 2010, fossil fuels underpinned Japan's energy supply structure, with nuclear energy and renewables playing a secondary role (McLellan et al., 2016; Okura et al., 2016). Pre-Fukushima accident, Japan had one of the most balanced power-generation portfolios among major power consumers, with no single fuel source accounting for more than one-third of its total generation capacity (EIA, 2017).

Like other industrialised countries, Japan is a major contributor to GHG emissions (Botzen et al., 2008). Since hosting the 1997 Kyoto Conference, Japan has established itself as a leader in climate policies with strong targets for GHG emissions and decarbonisation, and continued investment in energy efficiency technology and research (Aruga et al., 2024; Corbeau & Ledesma, 2016; EIA, 2017; IEA, 2019, 2023). Renewable energy capacity has increased in the past decade, despite geographical constraints around power generation potential (Aruga et al., 2024; Oshiro et al., 2016). Hydrogen is a major long-term focus for Japanese policymakers, with national targets focused on promoting hydrogen to support energy security and reduce environmental burdens. Hydrogen development is planned in three phases – promoting fuel cell electric vehicles, integrating hydrogen into energy system supply chains, and establishing a carbon-free hydrogen supply by 2040 (ACIL Allen Consulting, 2018; Jensterle et al., 2019; MFAT, 2023; Nagashima, 2018). The Government views hydrogen as a potentially lucrative new global industry.

Japan's energy mix

Japan has very limited natural resources. Between 2012 and 2017, on average, domestic fuel sources met less than 10% of the country's primary energy needs. Domestic natural gas production is declining, with reserves reaching approximately 3 BCM (BCF) in 2015. In 2022, domestic gas production rates were negligible, and 98% of gas demand was met through imported LNG (IEA, n.d.). Development plans for CH_4 hydrates – natural gas deposits trapped within crystallised ice structures – represented a potential source of domestic natural gas, but exploitation has been postponed because of high development and production costs (IEA, 2017). In 2016, Japan was the world's largest LNG importer and in the top four countries for imports of coal, petroleum, and other liquids, and for consumption of crude oil (EIA, 2017). Japan reduced its energy consumption by 35% between 1970 and 1990, and is targeting a similar improvement between 2012 and 2030 (EIA, 2017; IEA, 2021a). Indeed, Japan has the lowest energy intensity among all large developed economies resulting from a sustained and significant R&D investment in energy-efficient technologies since the 1970s (EIA, 2017; IEA, 2021a). In 2022, oil comprised the country's primary energy consumption by fuel type (39%), followed by coal (27%) and natural gas (21%).

Role of oil and coal

Oil demand has consistently decreased since the 1990s' recession because of Japanese foreign investment in industrial production in Asia, improved efficiency in the transport sector, and a switch to natural gas in the power sector (Taghizadeh-Hesary et al., 2017). In 2021, Japan had 22 refineries with a total of 3.7 barrels per day (b/d) of crude oil capacity. Reduced industrial output and reduced demand for oil products, however, have caused the sector to contract with excess refining capacity to manage (EIA, 2017, 2023; IEA, 2021a).

Coal is the primary fuel for power generation. In 2018, 61% of coal was used for power generation (IEA, 2021a). Coal's share in the sector rose from 23% before Fukushima to 31% in 2015 (Chapman & Itaoka, 2018; EIA, 2017; IEA, 2016b). Coal policy centred on three priorities – to secure stable low-cost supply, promote efficient power plants, and support exports of Japanese low-carbon technologies to countries continuing to use coal (IEA, 2016b; METI, 2015). As of April 2020, Japan had 150 coal-fired power plants, around 48 GW capacity, with a further 11 GW in the pipeline. The country's coal-fired infrastructure is among the youngest in the world; a coal phase-out would result in stranded assets of US$71 billion by 2030 (Trencher et al., 2020).

Role of renewables and nuclear

In 2024, Japan ranks sixth in renewable electricity production overall and third in solar power generation capacity (IRENA, 2024). The 2012 feed-in-tariff (FIT) helped increase renewable electricity generation from 10% of the total in 2012 to nearly 19% in 2019 (IEA, 2021a; Kucharski & Unesaki, 2018). Solar represents around 94% of renewable capacity growth since 2012, reaching 87 GW in 2023 (IRENA, 2024). Growth in renewables was factored into the 2018 SEP, with an ambitious target of 22–24% by 2030 (METI, 2018). Until the Fukushima accident,

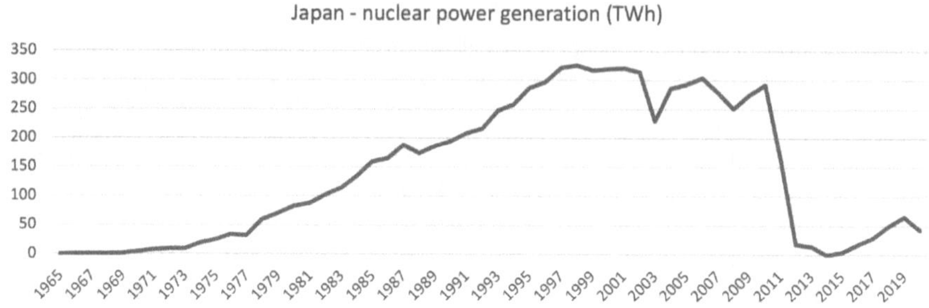

Figure 4.1 Japan's nuclear power generation (BP Statistical Review of World Energy, 2021)

Japan did not have ambitious renewable energy policies (Moe, 2012). Geographical challenges and grid limitations are obstacles for large-scale wind power – most wind resources are in the northern areas of Tohuku and Hokkaido, far from where most electricity is consumed (Oshiro et al., 2016).

The drivers and factors supporting green hydrogen development are currently unavailable. The Government's inability to reduce renewable energy costs consistent with global trends, an undeveloped grid network, and limited underground storage will prove challenging for power-to-gas, a process of converting renewable energy into hydrogen gas (Nagashima, 2018; Shibata, 2016). Figure 4.1 shows Japan's nuclear power generation, with a clear decline from 2011, the time of the Fukushima accident.

Carbon dioxide emissions

In 2018, the energy sector accounted for nearly 88% of total GHG emissions, followed by industry, agriculture, and waste management (IEA, 2021a). CO_2 emissions from fossil fuels approximated 1,181 million tonnes, 14% higher than 1990 levels, with power generation representing the largest CO_2-emitting sector with nearly half of all emissions. Natural gas-related emissions increased by 20% and coal by 15% between 2000 and 2022 (IEA, n.d.). After Fukushima, emissions in power generation increased by 54 million tonnes, mainly owing to the higher volumes of coal and natural gas used. Overall emissions dropped for five consecutive years to 0.99 billion tonnes in 2022 from a 2013 peak of 1.24 billion tonnes, a reduction of 20% (IEA, n.d.). Carbon emissions rose significantly in the immediate aftermath of Fukushima (EIA, 2017). Post-Fukushima, the GHG emissions-reduction target was revised to 3.8% by 2020 (based on 2005 levels), leading to an overall 3.1% increase over 1990 levels (IEA, 2016b). Japan exceeded its first target and commitment under the Kyoto Protocol by 2.4%, reducing GHG emissions by 8.4% in 2008–2012 from a 1990 base year (IEA, 2016b). There was no commitment made for the second period (2013–2030) but Japan pledged to reduce GHG emissions by 25% by 2020 from 1990 levels.

Natural gas dynamics and infrastructure

Japan has two power grids with insufficient interconnections and two incompatible power line frequencies of 50 Hz and 60 Hz, and is divided into ten regional power supply firms. Only 1.2 GW of electricity can be transferred between frequencies, severely limiting the possibility of developing an interconnected system (EIA, 2017; Oshiro et al., 2016). As of 2022, Japan operated 33 LNG terminals, with a total send-out capacity of approximately 290 BCM, which exceeded total demand, and an average utilisation rate of 37% in 2021 (EIA, 2023). Utilities, city gas companies, local governments, and large industrial companies own and operate LNG terminals, either alone or in partnership (IEA, 2016b). Most terminals are located near population and manufacturing hubs, such as Tokyo, Nagoya, and Osaka. Japan has an LNG storage capacity of around 18 mcm, which supports seasonal variations in demand and compensates for the lack of traditional gas storage. LNG storage capacity equates to approximately 36 days of domestic gas consumption (IEA, 2021a). Electricity companies own approximately half of the LNG storage capacity, while gas utilities own 40% (EIA, 2017; IEA, 2016b). In 2017, the total gas pipeline length was 261,167 km, of which 86% operated in low-pressure networks for local distribution and 13% in medium-pressure transmission lines (IEA, 2021a). There are no international cross-border pipelines, and the gas pipeline grid geographically covers only 5.7% of national land serving 65% of domestic gas demand, leaving liquified petroleum gas to meet the remaining demand. Overall, Japan's power and natural gas grids are underdeveloped and poorly integrated (IEA, 2016b, 2021a). The country's pipeline density is one-tenth that of Germany, with limited underground natural gas storage – just five small underground storage facilities and a total capacity of 1.4 BCM (IEA, 2021a; Shibata, 2016).

Japan was an LNG industry pioneer, beginning imports from Alaska in 1969 (EIA, 2017). The share of natural gas in the electricity supply mix increased from 2.4% to 29.3% between 1973 and 2000 (Colombo et al., 2016). LNG played a positive role in improving air quality, owing to lower SOX and NOX emissions (Corbeau & Ledesma, 2016). Reducing air pollution levels nationally was a key driver in the growth of gas use (Colombo et al., 2016) and natural gas consumption reached 4.4 TCF/year in 2015, rising by 42% from 2005 levels (EIA, 2017). Natural gas demand continued to increase between 2009 and 2012, plateauing in 2015 and then falling after that in response to lower economic demand, a drop in coal prices, and an increase in renewable energy use (EIA, 2017). Natural gas use for power generation grew significantly following the Fukushima accident (EIA, 2017). In 2019, the natural gas supply was 105 BCM, around 15% higher than in 2005, and 20% lower than in 2012 (IEA, 2021a). In 2014, natural gas as a share of total primary energy supply increased from 17.3% in 2010 to 23.3% in 2015 (IEA, 2016b). In 2019, natural gas accounted for around a quarter of total energy supply, and was the main fuel used in power generation with a 38% share (IEA, 2021a). In 2022, 62% of natural gas consumption was for power generation, followed by 19% for industry, 14% for residential and commercial, and around 3% for other purposes (IEA, 2023).

The Japan Gas Association demonstrated that up to 62 million tonnes of CO_2 can be reduced by 2030 by switching to natural gas in industry, fuel cell use, and cogeneration. The 2014 SEP (METI, 2014) solidified natural gas's future role, and the 2015 Long Term Energy Supply and Demand Outlook (METI, 2015) forecast that LNG import volumes would remain at approximately 62 MTPA by 2030 (these are worth contrasting to IEEFA, 2024, which forecasts that LNG imports will reduce by approximately 50%). The Ministry of Economy, Trade and Industry (METI)'s 2018 SEP promoted greater use of natural gas within industry and as a source for hydrogen production – both ways to support its GHG emissions-reduction objectives (METI, 2018).

Natural gas forecasts

A set of forecasts have predicted that overall primary energy consumption decreases, with lower-carbon scenarios anticipating more aggressive energy efficiency measures. Natural gas share and total amount of power generation drop significantly by 2050 in all scenarios except for the Institute of Energy Economics Japan (IEEJ)'s (2021) "reference scenario" that predicts a relatively stagnant demand for the next three decades (IEA, 2021b, 2023; IEEFA, 2024; IEEJ, 2021).

The future demand for natural gas depends on several key uncertainties, with the power sector driving potential changes, mainly around the role of coal use, short-term nuclear power ramp-up and long-term nuclear power share, and the ability for renewables to scale up rapidly and sustainably. LNG is likely to maintain its role despite increasing renewables and considering unclear nuclear pathways. LNG may be the best available solution given a paucity of options. It is unclear whether Japan will re-achieve 2010 nuclear energy levels in the future. There is a common view that both nuclear and coal have reached a ceiling in their consumption, and given the limitations of renewables, LNG is well positioned to continue acting as a key energy source in Japan.

The high cost of renewables is also seen as potentially slowing down aggressive growth. Thus, gas demand pressure was envisaged until the mid-2030s, with improvements to infrastructure, battery storage, and green hydrogen technology, and gas starting to come under increased pressure.

Key drivers and tensions

Interviews with key stakeholders by Al-Kuwari (2023) identified key variables and policy drivers affecting the future of natural gas and LNG. These included the "iron triangle" (polity, bureaucracy, industry), Fukushima and nuclear energy, security of supply, climate change and GHG emissions, and competitiveness.

The iron triangle

The interplay between industry, bureaucracy, and politics established a vested structure or "iron triangle" that plays a major role in determining the future energy mix (Moe, 2012, p. 262). The Japanese energy industry operates in a planned economy approach with the Government providing a "guiding hand" (Valentine & Sovacool, 2010). Historically, Japanese policymakers, with strong support from industry, favoured nuclear and other large-scale centralised thermal power technologies; this was reflected in the investment decisions made, eventually leading to institutional lock-in (Moe, 2012). This path dependency was instrumental in the fixed energy policy over the years, particularly around investment in low-carbon infrastructure, and was reinforced by the formation of special interest groups and the size and complexity of the country's energy systems (Moe, 2012; Okura et al., 2016). Deregulation, increasing public interest in climate change, and evolving relationship dynamics within government could break this inertia.

Evolution of the energy industry's role

During the Meiji era (1869–1912), before structured regulatory systems were introduced, private companies established their business practices, leading to the foundation of regional monopolies that controlled the electricity sector (Kucharski & Unesaki, 2018). In 1951, the American Occupation Authorities established a system that provided near monopolistic control over power generation, transmission, and distribution of electricity within exclusive regions to privately owned general electricity utilities (GEUs). These GEUs operated with little state intervention and regulation on the understanding that they would maintain a stable supply, increasing their influence over energy system decisions (Kucharski & Unesaki, 2018). Legislative mechanisms were introduced, giving the Government strategic control over technology decisions and resource planning, despite privatisation (Valentine & Sovacool, 2010), which served to strengthen the alignment of interests between industry and government. The concept of "coordinated capitalism", characterised by conformity in company organisation and the unique cross-shareholding structure between large companies highlighted by *keiretsu* relationships, also featured prominently in Japan's economy, encouraging coordination between stakeholders (Kucharski & Unesaki, 2018). From the 1960s, the Government actively supported Japanese companies through R&D funding, and imposed tariffs on strategic goods (Valentine & Sovacool, 2010). From 2014, Japanese companies participated in over 140 oil and gas projects, mainly in the Middle East, Southeast Asia, Australia and, more recently, North America (EIA, 2017).

Ten privately owned integrated power companies act as regional monopolies and account for approximately 80% of Japan's installed power generation capacity (EIA, 2017; IEA, 2021a). Utility companies and the nuclear industry are the main participants in the energy system and extend their influence through Denjiren (the Federation of Electric Power Companies of Japan). Historically, they have been successful in blocking large-scale efforts to establish rival power-generating companies (Moe, 2012). However, within the energy and gas industries, there have been strategic changes among companies to evolve beyond traditional models in preparation for deregulation, and to adapt to energy-transition-related pressures. For instance, Chubu Electric, a regional GEU, was the first company to go against established norms and compete outside its home market of Nagoya – competing with Osaka Gas and TEPCO. TEPCO and Chubu Electric created a joint venture, JERA, in 2015, and Tokyo Gas and Kansai Electric subsequently formed a similar LNG procurement partnership. Energy companies have also worked with corporates (e.g. Softbank and NTT Docomo) to provide other wide-ranging and bundled products and services to increase their market share (EIA, 2017; Kucharski & Unesaki, 2018; Trencher et al., 2020).

Politics and bureaucracy

After the Second World War, Japan's political system was restructured to prevent a return to authoritarianism (Moe, 2012). Consequently, a relatively weak executive branch of politicians and political appointees was supported by an experienced and well-staffed administration. This created a system in which bureaucrats and industry decision-makers could dedicate resources to promote consistent energy policies, even as governments changed (Moe, 2012). Japan's energy policymaking structure therefore remained stable for four decades and led to a predictable and consistent energy policy and direction. The political arena was almost exclusively controlled by one party, the administration was dominated by METI, and business interests were represented by industry. The METI-hosted Shingikai (advisory committees) included members representing the interests of the energy industry, and while this supported strategic alignment, it also reinforced the interests of those groups to the detriment of any changes favoured by NGOs or renewable industry companies. An example of these close links is the civil service's early retirement system, which gives early retirees positions within the energy industry. This can lead to bureaucrats working for companies and sectors they once regulated, further reinforcing interinstitutional links (Moe, 2012; Trencher et al., 2020). This slowly began to change after the Fukushima disaster (Kucharski & Unesaki, 2018).

The Democratic Party of Japan (DPJ) ousted the Liberal Democratic Party (LDP) from power between 2009 and 2012. However, the DPJ lost the 2012 general election mainly because of perceived mismanagement in the aftermath of the Fukushima accident (Efird et al., 2018), and through resistance from METI and industry. The DPJ unsuccessfully formed a climate inter-ministerial task force to improve the energy policymaking process; it politicised the debate around future energy policy post-Fukushima by setting up the Energy and Environment Council to investigate costs and attempting to shift the blame for the accident to the utilities (Efird et al., 2018; Moe, 2012; Nishimura, 2016). The LDP Government later reversed these DPJ decisions and even recategorised nuclear energy as "green", a move that epitomised the powerful role bureaucrats, and METI in particular, played in shaping energy policy.

The delayed low-carbon transition

The renewable energy industry did not gain strong support from policymakers, despite Japan being active in wind and solar technology development since the

1970s (Cherp et al., 2018). Coalitions of emerging renewable companies were unable to secure stronger policies, while the nuclear sector strengthened politically because of its continued growth and its role in meeting growing electricity demand (Cherp et al., 2018). The solar industry was more successful in lobbying METI and the administration by positioning itself, with METI's support, as an emerging high-tech direct-to-consumer export industry. The introduction of solar power did not require major structural changes to the energy system, and it leveraged Japan's manufacturing and engineering expertise while supporting overall industrial export levels. Wind energy, conversely, challenged Japan's energy structure because it competed directly with large-scale energy supply controlled by utilities and independent power producers. Japan's geography and grid challenges also meant that increased investment was needed to move wind-generated electricity from the north to the rest of the country (IEA, 2016b). The support of clean coal showed how industry and the Government could influence public opinion. The claim was that Japan's superior technological ability could develop a coal use industry compatible with climate change – domestically and internationally (Trencher et al., 2020).

Debates on the future energy mix call into question the strong relationship established between the three "iron triangle" participants. Increased public and media interest has expanded the scope of energy debates and led to growing misalignment in the ability to deliver ambitious climate goals. Similarly, while the Government favours a nuclear restart – with METI supporting ramp-up to reach pre-Fukushima levels as a minimum in the coming years – there is scepticism about whether this can be achieved and that government plans are not private sector friendly. Important decisions may have been taken without business interests in mind, or even in consultation with industry – a potential break from previous practice.

Issues within the administration

Environmental issues have clashed with industry and business sector objectives (Okura et al., 2016) in Japan. The Ministry of Energy (MoE) often pushed for more aggressive environmental measures to be incorporated into energy policy but had limited influence compared with METI. The different views of the two ministries highlight the current challenges in the country. METI's energy policy drivers focused on industrial and economic development, not climate change mitigation. Meanwhile, the MoE, supported by increased public awareness of environmental issues due in part to the Fukushima accident and coupled with stronger leadership, pushed METI to consider the environment in its energy policymaking. METI has been described by some as the "accelerator" and the MoE as the "brake", implying that energy transition and climate considerations slow development. METIs role may be seen as stronger, pointing to the MoE's lack of legal authority to enact changes. Some industry participants were explicit when asked about the influence of both ministries and concluded that METI is more important than MoE (Al-Kuwari, 2023). Many utility companies (e.g. Tokyo Gas, TEPCO, Osaka Gas, and JERA) follow the policy of METI because it is larger than MoE and has more power. The popular MoE minister Shinjirō Koizumi boosted the environmental agenda and helped it gain momentum through energy discussions with METI that led to a shutdown of older coal-fired power plants. The MoE is also supported by the media's increased focus on GHG emissions and support for environmental matters. Increased alignment between METI and MoE could serve to elevate environmental considerations in energy policy.

Impact on LNG

The "iron triangle", partially responsible for decades of stable policies, is weakening and potentially changing in shape and structure. Industrial players consider that the Government's climate policies are too ambitious with the bureaucracy internally balancing competing economic and environmental objectives. The increased deregulation and a reduction of cross-shareholding impedes the industry's ability to advance in a coordinated manner during a time of transition. Furthermore, the public's role is gaining importance by demanding greater transparency and debate on energy-related matters, particularly nuclear energy. With the support of the media, and a growing civil society, the decades-long system that established path dependency is weakening.

LNG could play a continued role as part of the evolving structural changes in Japan's energy mix. From a policymaker's perspective, it is a flexible and acceptable interim option that benefits from an increased global supply and measures taken to improve competitiveness and has environmental benefits over coal. LNG infrastructure may be used to provide support for any significant shifts in industry away from nuclear and coal. From a public and environmental perspective, LNG supports the development of renewables and hydrogen economies.

Fukushima and nuclear energy

Nuclear energy was introduced under the 1955 Atomic Energy Basic Law, which limited the use and purpose of nuclear technology to peaceful purposes related to securing energy resources to meet the needs of Japanese society (IEA, 2016b, 2021a). After the oil shocks of the early 1970s, Japan embarked on a nuclear power development programme focused on light-water reactors (LWRs) using enriched uranium in pressurised- and boiling-water reactors (IEA, 2016b). The first nuclear plants were based on US designs and built-in partnership with Japanese companies, which then constructed additional plants under licence. The Government committed over US$71 billion per year between 1975 and 2001, including most of its energy R&D spending, to develop the world's leading nuclear industry. Government support led to the development of three major power plant manufacturers (Hitachi, Toshiba, and Mitsubishi) and the largest number of nuclear researchers, scientists, and engineers per capita (IEA, 2016b, 2021a; Valentine & Sovacool, 2010). By the end of the 1970s, Japanese companies had proven their nuclear capabilities, constructing Generation III reactors faster and more effectively than their US and European counterparts. In the 1980s, Japan developed an energy mix centred on increasing nuclear power while reducing energy intensity and increasing investment in renewable energy R&D (Cherp et al., 2018).

Nuclear energy also supported Japan's foreign affairs political agenda through knowledge export and by underpinning technology development and research alliances with the United States and Europe. The combination of nuclear technology know-how and strong anti-proliferation policies positioned Japan as an advocate for peaceful nuclear energy use, elevating its international profile (Len & Nian, 2014). Japan leveraged its nuclear capabilities to support its long-term energy security objectives by entering into transfer deals where crude oil supply contracts were exchanged for technology transfers (EIA, 2017). The industry's strategic and national security importance was bolstered by the ability to manufacture key components for US and French nuclear reactors, and the belief among the ruling elite that nuclear technology capability should be maintained as a deterrent (Len & Nian, 2014).

Support for nuclear energy strengthened within METI, as nuclear-related exports became the focus of the "new green growth policy" (Moe, 2012). Nuclear was also a main pillar of Japan's 2006 New Energy Strategy that aimed for 40% of power generation from nuclear by 2030 because of its carbon reduction potential (Valentine & Sovacool, 2010). METI's 2010 SEP called for an increase in nuclear energy, with its power generation share to reach 53% by 2030 compared with 25% in 2007. By early 2011, Japan was the world's third-largest producer of nuclear power; its power generation mix included 26% nuclear energy (EIA, 2017; IEA, 2021a; World Nuclear Association, 2024). In the early 2010s, the nuclear energy industry was optimistic, with the IEA forecasting that global nuclear capacity would increase from 370 GW in 2009 to 1,200 GW by 2050 (Hayashi & Hughes, 2013). Despite safety-related challenges, nuclear energy was "promoted as a necessary evil for supporting economic prosperity" (Valentine & Sovacool, 2010, p. 7974), with studies concluding that a 100% nuclear energy-based electricity system was economically and technically feasible by the end of the century with the support of hydrogen and storage technology (Zhang et al., 2010).

Fukushima impact

On 11 March 2011, a 9.0-magnitude earthquake hit off the coast of Sendai, triggering a tsunami that damaged the Fukushima Daiichi nuclear plant and led to the immediate shutdown of 10 GWs of electricity capacity (EIA, 2017). In the summer of 2011, users across Japan adopted power-saving measures to manage supply shortages and for the next two years, nuclear power generation was gradually suspended for the first time in 40 years (EIA, 2017; IEA, 2016b, 2021a; Nishimura, 2016). The first nuclear restarts took place in the second half of 2015 (IEA, 2021a) when utility companies received approvals to recommission some reactors, although continued public scrutiny and opposition contributed to uncertainty about the ramping up of additional nuclear capacity. All six Fukushima Daiichi nuclear plant reactors, a combined capacity of 4.6 GW, were decommissioned, with another 2.6 GW of smaller, older reactors also decommissioned by

2017 (EIA, 2017). As of June 2024, there are 33 operable nuclear reactors, with 12 approved for restart and a further 13 in the process of restart approvals (IEA, 2021a; World Nuclear Association, 2024). The share of nuclear in the total primary energy supply in 2022 was 3.7%, down from 15.0% in 2010 (IEA, n.d.).

Institutional reforms: Structural and regulatory

The Fukushima accident exposed technical and regulatory problems with the electricity sector, including increases in electricity costs and power outages, partly owing to the inability of regional monopolies to transmit outside their regions (EIA, 2017). In the aftermath of the accident, electricity reforms focused on securing stable electricity supply, expanding consumer choice, and reducing electricity rates (IEA, 2016b; Kucharski & Unesaki, 2018). The Nuclear Regulation Authority (NRA) was established under the MoE as an independent nuclear regulator to remove any potential conflict of interest with METI (Efird et al., 2018; IEA, 2021a; Kucharski & Unesaki, 2018). Local governments had to approve the start-up of compliant or upgraded nuclear plants through the NRA. Decisions related to nuclear start-up were therefore transferred to regulators and local officials (EIA, 2017; Kucharski & Unesaki, 2018), giving the public the ability to influence energy policy decisions.

The safety risks related to earthquakes will remain a consideration in the future of nuclear power, particularly considering the greater tsunami potential along the Pacific coast where many facilities exist. The uneven geographical distribution of nuclear power plants and lack of suitable regional interconnections are also important factors in those scenarios where certain facilities do restart (e.g. around Tokyo, Chubu, and Tohoku) (McLellan et al., 2013). The Fukushima accident will affect long-term nuclear energy affordability given the additional safety measures imposed and the higher costs associated with extended construction periods (Hayashi & Hughes, 2013).

Energy mix

Pre-Fukushima, Japan's efforts to achieve a more self-reliant lower-carbon energy system were heavily centred around an increased role for nuclear power. The accident had a deep impact on policymakers, requiring Japan to rewrite such policies and develop a vision for nuclear power plants to tackle global climate change and balance economic growth for Japanese industries. Released several years after the Fukushima accident, the 2014 SEP still promoted nuclear power as an important source and a "quasi-domestic important low-carbon baseload power source, contributing to stable supply demand as long as safety can be ensured" (METI, 2014, p. 24). The plan also highlighted how dependence on nuclear power will be reduced through increased energy efficiency measures and renewable energy supplies (EIA, 2017; IEA, 2016b). METI's 2018 SEP main policy direction for nuclear energy prioritised safety and looked to renewable energy for decarbonised electricity in the long term, while still encouraging nuclear power based on its proven stability as a baseload energy source, its historically low and stable operational cost, and its low-carbon nature (IEA, 2021a; METI, 2018). In recent decades, government officials have avoided taking a position on nuclear power for political reasons. Fukushima, however, brought these issues into the spotlight and people became more aware of nuclear policy. While this increased knowledge highlighted some of the benefits of nuclear energy, it generally led to negative sentiment that may affect future energy-related debates.

During its brief period in power, the DPJ adopted an ambitious environmental and energy-mix target in the 2010 Basic Energy Plan. This included increasing the share of renewables and nuclear power and raising the low-carbon emission ratio to 70% by 2030, resulting in a 30% reduction in energy-related CO_2 emissions from 1990 levels (Len & Nian, 2014). The DPJ was forced to revise the plan post-Fukushima, however, and pledged to shut down all nuclear power plants by the end of the 2030s (Len & Nian, 2014). The Government initially had public and media support and promoted the concept that nuclear energy could be feasibly and economically replaced by renewables, supported by the fact that consumers did not experience price changes owing to fragmented liberalisation and regulation protection (Nishimura, 2016). Thus, the debate around future nuclear policy was largely focused on safety and did not consider the wider benefits to the country. Despite ongoing anti-nuclear resistance from the public and some government factions at the time, Prime Minister Abe's LDP Government reversed the DPJ's closure plans in 2012. One energy industry participant confirmed that Japanese utilities and industry had been pushing hard for renewed commitment to nuclear capacity, and that utilities are keen to restart nuclear power, strongly supporting the Government in this respect (Al-Kuwari, 2023).

Impact on LNG

LNG was the fuel of choice to make up for lost nuclear power generation after Fukushima (EIA, 2017; Hugo & Kusumawardhana, 2023; IEA, 2021a). The LNG market's flexibility led to it meeting very high short-term demand with significant volumes. Japan's LNG imports increased from 98 BCM in 2010 to a high of 123 BCM in 2014 (reducing to 92 BCM in 2022) (EIA, 2023), increasing the overall thermal power generation to nearly 90%, with LNG accounting for 45% of total power generation (Colombo et al., 2016). In 2014, construction began on 12 new gas-fired power generation plants in anticipation of a continued reduction in nuclear power and to replace the more expensive oil-fired plants (Len & Nian, 2014). However, increasing LNG imports came at a cost (Boersma & Jordaan, 2017). Those buyers with flexibility reprioritised importation of oil and coal products, while those fully dependent on gas-fired power generation continued to import LNG regardless of cost, which reached US$18/MMBtu in 2012 (Colombo et al., 2016; IEA, 2016b).

Interviews with stakeholders revealed a general recognition that nuclear energy will continue to play an important role in the energy system (Al-Kuwari, 2023). At the same time, a common view highlighted waning political support, with politicians more reluctant to publicly support nuclear power. Divisions within the LDP were also made apparent, weakening long-term support for nuclear with an internal struggle in the party on how to manage this delicate issue. An energy expert observed that the impact of Fukushima is still being felt many years later and that the pace of restarts will still factor into Japan's LNG needs through the 2020s. With more clarity around the trajectory of renewables and coal use, LNG may be the fuel most directly affected by changes in the role of nuclear power.

In terms of Japan's future energy mix, companies must plan for two potential pathways – one with a 2018 SEP-aligned target and one with little or no nuclear power. The industry has accepted that it is not realistic to assume that nuclear energy use will recover to pre-Fukushima levels. Industry executives noted that long-standing uncertainty around nuclear power could add to the overall cost to energy systems and coupled with negative public opinion, could facilitate the transition by making it easier to secure investment support for renewables and hydrogen over maintaining or upgrading existing nuclear power plants (Al-Kuwari, 2023). While nuclear power will continue to play an important geopolitical, economic, and industrial role in Japan, that role is unlikely to return to pre-Fukushima levels. Limitations on coal development will enable renewables and LNG to grow, with LNG able to leverage existing capacity to scale up more rapidly. Natural gas could be considered the next best option after nuclear energy in Japan in the coming decades, taking into consideration competitiveness, energy security, and the environment.

Security of supply

With virtually no oil and rapidly declining and negligible natural gas reserves and coal production ending in 2002, Japan's economy relies heavily on imports for energy production and industrial raw materials. The country imports almost 100% of its fossil fuel needs and had a low self-sufficiency ratio of 12% in 2019, placing it 35th among OECD countries (Agency for Natural Resources and Energy Japan, 2019; EIA, 2017; IEA, 2021a; METI, 2022). Self-sufficiency gradually improved pre-Fukushima and remained consistent at approximately 80% based on energy efficiency efforts, increased nuclear energy production, and, increasingly, renewables. Post-Fukushima, however, import dependency rose to 94% of primary energy supply in 2013, compared with a consistent 80% in the first decade of the 2000s (Agency for Natural Resources and Energy Japan, 2019). After the accident, policymakers needed to consider how to address energy security even more urgently given the uncertain nuclear capacity ramp-up and decarbonisation agenda. Japan had been badly affected by OPEC's 1973 oil embargo, leading it to prioritise energy security objectives and diversify resources (Colombo et al., 2016; Solomon & Krishna, 2011). Coupled with an increased electricity demand, the country pursued increased nuclear power capacity in the 1970s (Cherp et al., 2018; IEA, 2016b). Energy planners focused on reducing reliance on oil – which had represented 77% of the energy mix in 1973 – while meeting energy-intensive industrial demand, resulting in a greater proportion of coal, gas, and nuclear. Japanese companies also pursued secure supplies by financing, engineering, constructing, and project managing international large-scale projects (EIA, 2017).

In the 2014 SEP, "energy security" represented one of the "3 Es", alongside economic efficiency, and environmental protection. The 2015 Long Term Energy Supply and Demand Outlook set out a vision for a future energy supply-and-demand structure with increasing self-sufficiency. Meanwhile, the updated "sophisticated 3E+S" in the 2018 SEP introduced plans to strengthen energy security based on diversification of energy options and greater self-sufficiency (IEA, 2016b, 2021a; METI, 2015, 2018, 2022). Regardless of emissions targets and climate change politics, energy security remains a priority for Japanese policymakers and will continue to shape future energy and environmental policy. Despite the country's decarbonisation ambitions to ensure national energy security, the stable import of energy resources remains critical. From an energy security perspective, there are two categories – imports (i.e. fossil fuels) and domestic resources (i.e. renewables, nuclear energy, and to an extent energy efficiency). Hydrogen, which requires its own investigation, plays a unique role in that it can serve as a bridge to reduced import dependency.

Since the 1970s, a diversified energy mix has been viewed as the best mitigation against supply shocks, cementing coal's role. Highlighted in the 2018 SEP as an important energy-mix component, coal's future role is still justified by the energy security lens (IEA, 2021a; METI, 2018; Trencher et al., 2020). Coal's role in the future energy mix is under pressure from long-term emissions goals, competition from increased LNG supplies, and a growing renewables portfolio. However, a growth in coal-to-gas switching increases reliance on LNG, which carries economic and energy security risks, while future market liberalisation could make coal-fired generation more economical (Arima et al., 2015). Japan has historically adopted an "all of the above" approach to ensure a role for coal in the medium term despite climate concerns. Coal-importing companies will continue to explore ways to utilise their coal assets by integrating ammonia or other hydrogen carriers into their plans, part of a more aggressive approach from utilities to address the transition, under the guise of maintaining a diversified but environmentally responsible portfolio. The ongoing support for clean coal technology development could mean that coal continues to have a role, with the ability to export knowledge and technology to new markets. METI's announcement to retire older, inefficient coal plants and the Government's support for ceasing investment in coal technology overseas could represent a great change in the energy mix going forward (Al-Kuwari, 2023). Coal policy will be a focus over the next decade in terms of the evolution of fossil fuels in Japan, supporting the notion that LNG demand could grow if Japan does not approve a significant amount of new coal-fired power stations.

Oil will continue to have a role in the energy system for decades with limited ability to switch to alternatives in the short to medium term where transport and industrial demand dominate. Lowering demand and increasing energy efficiency and diversification are key mitigating tools for switching to LNG, with some potential for fuel oil and diesel.

Low-carbon electricity and self-sufficiency

A balanced and distributed renewable energy portfolio can enhance energy security and improve resilience against natural disasters, while supporting emissions-reduction targets. Given that nuclear power still requires some level of international involvement, the largest share possible of renewable energy would be necessary to support energy security goals (McLellan et al., 2013). The solar industry has been influential within METI since the 1970s (Moe, 2012). In 1974, METI launched the "Sunshine Project", a 25-year solar energy technology R&D plan in response to the first oil shock (Zhu et al., 2020). The IEA described renewable energy in Japan as "both a strategic opportunity and a practical challenge" (IEA, 2016b, p. 12). These challenges were laid out in the 2018 SEP, which set a renewables target of 22–24% by 2030, ambitious for Japan but not sufficiently high to mitigate potential delays in the ramp-up of nuclear energy or replace fossil fuels in the near to medium term. A Bloomberg (2020) Energy Finance report concluded that Japan is one of 39 countries unable to generate 50% of electricity from renewables by 2050 given its limited land availability. While increasing renewables improves self-sufficiency, the industry considers it expensive and unable to fully replace the stability provided by fossil fuels within the current system. Government targets represent the best-case outcome for renewables, and – at least for the time being – the most realistic solution to meeting any major demand gaps is to use natural gas (Al-Kuwari, 2023).

Nuclear power still plays a crucial future role in determining pathways for energy security and transition. McLellan et al. (2013) evaluated scenarios with aggressive renewable energy penetration and nuclear energy, concluding that even with renewables reaching 35%, well above the 2014 SEP target, energy dependence would decrease to approximately 85%, that is, near pre-Fukushima levels. Renewable energy potential cannot eliminate the need for fossil fuels to 2030 and beyond, striking a significant blow to energy security efforts in the absence of nuclear power. Arima et al. (2015) concluded that nuclear start-up and ramp-up are prerequisites for Japan to achieve cost and emissions reductions while improving energy security. Oshiro et al. (2016) predicted that by 2030 import dependency will rise to levels higher than in 2010 if nuclear plants are not reinstated.

The lack of excess or low-cost renewable energy reduces Japan's ability to produce hydrogen from electrolysis, or green hydrogen, in the short to medium term (Nagashima, 2018), leaving the country dependent on either importing green hydrogen or producing blue hydrogen from imported fossil fuels with CCS. Large-scale CCS in Japan is limited and expensive. Thus, CCS is conducted abroad, or CO_2 is exported, further complicating the supply chain and overall economics. Ishimoto et al. (2017) projected that a global market of CO_2-free hydrogen by 2050, supported by liquified hydrogen transportation and reduced production costs, could have a positive impact on Japan's energy system (Ishimoto et al., 2017). Japan's hydrogen plans could bolster LNG's role in the short term, although it is unclear how the pathways will develop in the absence of domestic CCS. Hydrogen, however, was not considered in the short-term plans of the Japanese LNG importers interviewed

(Al-Kuwari, 2023). LNG's role will be bolstered by a hydrogen future – natural gas should be used to produce hydrogen (blue hydrogen) from 2030 to 2040 to bridge between a natural gas and a hydrogen economy.

Resource diplomacy

METI and Japanese policymakers have historically supported fossil fuel importers politically to facilitate access to a wide range of resources. Active resource diplomacy and financial support for upstream investment have been part of the Government's energy policy for decades (IEA, 2016b). The 2018 SEP prioritised supporting Japanese to participate in overseas upstream and development projects, considering this key to securing resources, like China and India have done (Agency for Natural Resources and Energy Japan, 2019). Resource diplomacy relationships differ by fuel – with the UAE and Saudi Arabia for oil, Australia and Qatar for natural gas, and Australia and Indonesia for coal. In 2022, Australia was the largest LNG exporter to Japan at 429% of total volume, followed by Malaysia and Russia at 17% and 9%, respectively (IEA, 2021a, 2023). Policy planners have long viewed LNG as an important solution to energy source diversification. Over time, the focus has shifted from maintaining strong relations with gas-supplying countries to developing market-driven instruments to secure supplies (Colombo et al., 2016; METI, 2018; Stern, 2016).

Growing the LNG market

Japan has succeeded in diversifying its LNG imports over time (IEA, 2021a, 2023; Vivoda, 2019). The addition of LNG exporters, including Russia, the United States, and Middle Eastern countries, increased supply options. LNG imports from the United States specifically supported energy security goals. Firstly, they opened additional supply routes by sea to Japan, beyond the existing South China Sea route that represented up to three-quarters of Japanese imports and was a significant geopolitical risk. Secondly, increased trade in energy could strengthen the security partnerships that support Japan's geopolitical regional role in Asia in an era of increased Chinese energy market influence. Thirdly, US LNG export projects offered contracts indexed to Henry Hub prices with no obligation to take cargoes when prices were high and with full destination flexibility. This increased interest from Japanese utilities and trading houses to invest in US projects and consequently import significant volumes, representing up to 25% of total LNG imports (Colombo et al., 2016). Japanese companies also sought to expand imports from Russia, Mozambique, and Canada, while maintaining a secure and stable supply from the Middle East. Qatar is seen as crucial for its ability to divert large volumes destined for the United States and Europe, as witnessed in the immediate aftermath of Fukushima. For other Asian countries, including China and India, the goal would be to cooperate as importers to tackle common structural issues related to LNG pricing and other market dynamics (Colombo et al., 2016; METI, 2018; Stern, 2016). Energy security is crucial to government policy and a critical element for utilities and energy suppliers. Although there is no obligation for companies to maintain emergency stocks of LNG or other fuels, utilities and gas companies maintain a month's supply as a standard practice (IEA, 2016b). LNG could, therefore, increase its energy-mix share from a diversification perspective.

Increased flexibility

LNG could play a supportive role in enhancing Japan's security of supply, although increased reliance on LNG to replace nuclear energy increases overall fossil fuel imports and reduces self-dependency. Japan could take a global approach to the LNG market and develop a liquid trading hub to stabilise supply for buyers and reduce volume risk to sellers, with the overall objective of establishing more accurate supply and demand-based pricing. Using spare capacity coupled with improved market-based flexibility supports the use of LNG as a tool for managing short- and medium-term supply-and-demand changes. Increased and diversified global supplies and a gradual increase in blue hydrogen (produced via LNG imports to develop the necessary hydrogen infrastructure) could support long-term security-related goals.

Climate change and GHG emissions

Japan hosted the UNFCCC's third formal meeting in 1997 where negotiations for the world's first international treaty to address emissions reductions took place – the Kyoto Protocol (CarbonBrief, 2018; IEA, 2016b). Japan met its Kyoto Protocol first commitment period (2008–2012) and reduced GHG emissions by 8.4% from a 1990 base year – 2.4% above its target of 6% – achieved by using carbon sinks and international flexibility mechanisms (IEA, 2016b). At the United Nation's COP15, Japan did not commit to Kyoto's second period (2013–2030) but pledged a reduction of 25% in GHG emissions from 1990 levels by 2020. This target was to be met by increasing low-carbon generation and growing the combined share of nuclear and renewable energy within the energy mix to 70% by 2020 compared with 34% in 2007 (IEA, 2016b). The 2010 SEP indicated an increase from 30% to 50% for nuclear energy alone, making nuclear energy the central mechanism for reducing emissions (METI, 2010). In 2013, in the aftermath of the Fukushima accident and under the planning assumption that nuclear power would not restart in the short term, Japan set a target of a 3.8% reduction in GHG emissions to 2020 below 2005 levels, which translated to a 3.1% increase in emissions over 1990 levels (IEA, 2016b). In December 2015, Japan announced a 2013–2030 reduction of 26% in its intended nationally determined contribution (INDC) during COP21, a reduced target compared with that announced in 2009 in Copenhagen (IEA, 2016b; Nagashima, 2018). This pledge represents an 18% reduction from 1990, below Europe's 40% reduction, by 2030, a commitment criticised by the climate community for its lack of ambition (CarbonBrief, 2018).

Priorities and pressure

In May 2016, the Government released the Plan for Global Warming Countermeasures, which set out the country's path to achieving the 2030 INDC targets and the 2050 strategic goal (IEA, 2016b). The conditions underpinning the plan included compatibility with economic growth and making use of the most economical technology to achieve CO_2 reductions. A long-term challenge for Japan continues to be how to reduce 80% of GHG emissions by 2050, while simultaneously growing the economy and addressing climate change (Tagami, 2020). Recently, more aggressive commitments have been integrated into the country's energy and environmental policies. The MoE published its "Long-Term, Low-Carbon Vision" in 2016, which set a mandate beyond reducing GHG emissions by 80% by 2050 to deal with the economic and social issues caused by climate change (Ministry of Environment, 2017). In October 2020, Prime Minister Yoshihide Suga announced a 2050 net-zero target with reduced reliance on coal and increased use of renewables and nuclear energy (CarbonBrief, 2018; IEA, 2021a; MOFA, 2021).

Japan's initial involvement in supporting the Kyoto Protocol and its classification as a leader affects its international profile. Geographically and policy-wise, the country faces little regional pressure on how emissions agreements are implemented compared with other economies because it is not part of a highly integrated set bloc of countries or tied to a regional framework. Although Japan is a party to international climate change agreements, it will be up to the Government to respond based on domestic implementation needs in the absence of overriding regional pressures.

Energy mix and emissions trajectory – "unprecedented uncertainty"

Japan's energy system needs significant restructuring to meet climate objectives. In 2018, despite the restart of some nuclear capacity and increased renewables use, approximately 80% of Japan's electricity still depended on thermal power (Agency for Natural Resources and Energy Japan, 2019). The 2018 SEP "3E+S" strategy set out plans for how GHG emissions-reduction commitments would be met with reduced nuclear energy – by focusing on energy-saving measures, increased renewable energy capacity, and improvements in thermal power-generation efficiency (IEA 2021a; METI, 2018).

METI (2015) conducted a sensitivity analysis to calculate the impact of a combination of energy-mix compositions on CO_2 emissions. The analysis found that higher LNG use would be beneficial if replacing coal, but adverse if replacing nuclear power or renewables. Japan's marginal abatement costs for CO_2, and its CO_2 emissions-reduction cost as a measure of GDP are significantly higher than for the EU, Russia, China, and the United States, making the challenge, and therefore climate ambitions, more difficult by comparison (Arima et al., 2015). METI and the MoE encourage different approaches to long-term emissions reductions. METI generally promotes self-regulation and voluntary approaches to climate issues, supported by international initiatives to reduce emissions. The MoE, with the support of the Ministry of Foreign Affairs, views the carbon pricing system as a central component to climate strategy, while METI is critical of its potential impact on the economy (Nagashima, 2018; Trencher et al., 2020). The increased awareness of environmental considerations has improved the alignment between MoE and METI, who are now in the same position in terms of balancing and tackling global climate issues and supporting economic recovery. In the past, in contrast, the two issues were treated as different, sometimes competing pillars. There has also been a shift with METI recognising the need to tackle global climate change.

Challenges: Nuclear, coal, and renewables – "no magic bullet"

Pre-Fukushima projections of Japan's energy mix concluded that policies to reduce GHG emissions would lead to a decline primarily from coal and oil, followed by natural gas and then nuclear, with renewables playing a larger role in the future (Valentine et al., 2011). Hong et al. (2013) analysed several future pathways for Japan's energy system, applying different proportions of nuclear energy, including a pathway with no nuclear at all. The authors concluded that the pathways with the greater nuclear share were the most beneficial from economic, environmental, and social perspectives. Addressing the dichotomy characterising some environmentalists, Arima et al. (2015, p. 19) noted:

> [I]n Japan, those who are against new coal-fired power projects often also oppose the restart of nuclear power plants. The best way to minimise the entry of new coal-fired power plants is, however, to ensure the steady reoperation of nuclear power plants.

In 2020, after years of pressure from the MoE, METI announced the gradual retirement of older, inefficient coal-fired plants domestically and the end of governmental investment in coal technology internationally (EIA, 2023). Renewables are the key component in Japan's low-carbon ambitions because they can meet the most domestic demand while addressing other priorities. Long term, hydrogen could become the biggest emissions lever for Japan, a contention that "rests on the firm belief that hydrogen can be a decisive response to [the country's] energy and climate challenges" (Nagashima, 2018, p. 5). The main criterion for successful homegrown green hydrogen production capability is the sustained and economic growth of renewable energy.

LNG – part of the solution?

A stable supply of LNG would support sustainability, stable livelihoods and economic activities, and solve the issue of climate change. The view that LNG can play a positive role in Japan's energy transition was widely shared by Al-Kuwari's (2023) industry, energy expert, and policymaker interviewees. Statements related to LNG and its role in emissions included: "We need LNG for decarbonisation"; "LNG is part of the environmental solution"; "LNG is already playing a very important role to facilitate decarbonisation in Japan"; and "We believe natural gas is the important energy source for climate change and economic growth and stable supply". Although natural gas will eventually face difficulties, being a fossil fuel, the situation in Japan differs from that in Europe and there is still scope for LNG to play a positive role in balancing economic and environmental goals.

More measures can be taken to support LNG in Japan, and buyers and sellers could work together to lay some of the groundwork. Carbon-neutral cargoes and similar schemes are new phenomena, and it is unclear whether they will play a significant role in both sellers' and buyers' portfolios – indicative of the innovation and experimentation that LNG participants in Japan are willing to consider. There are challenges with carbon-neutral cargoes because the extra premium cannot be transferred to customers until it is accepted, or the carbon accounted for with a value from a regulatory perspective. Thus, its effect will be limited to public relations purposes. This practice is partly in response to demand from a certain segment of customers – eco-conscious companies in hospitality and cosmetics, for instance. As end-users' awareness of the role LNG plays increases, LNG importers may come under greater pressure. Some of the main Japanese importers are moving towards a hydrogen solution instead of focusing on low-carbon LNG, which could particularly impact LNG producers if there is a move towards green hydrogen.

Competitiveness

Reasons for liberalisation and deregulation

Electricity and gas deregulation efforts were introduced to stabilise electricity supply, expand consumer choice, and reduce electricity rates (IEA, 2016b). The Government's focus on ensuring security of supply contributed to development of an efficient structure in which state-owned buyers of imported fuels could transfer procurement costs to consumers, disconnected to price developments in international markets (Tusiani & Shearer, 2016). Specifically, the absence of an independent system operator and lack of separation between transmission and generation in the retail sector meant that fluctuations in supply were not reflected in retail prices (EIA, 2017; Nishimura, 2016). By the time of the Fukushima accident, however, major structural issues persisted in the electricity sector, including the inability of regional monopolies to transmit outside their regions, leading to higher costs and power outages. Utilities could also not raise the price of electricity immediately following the accident, despite supply disruptions, increasing the price of fuel and generation. Instead, a lengthy months-long METI-managed authorisation process was required to revise regulated tariffs (Nishimura, 2016).

Moving towards a decentralised lower-carbon system supports the country's sustainability goals and improves the electricity system's reliability based on greater flexibility and an ability for small systems to better respond to shocks (McLellan et al., 2016). To address these structural issues, a three-phase energy market liberalisation plan was launched in 2016. Phase one was the liberalisation of the retail electricity market, where start-ups and smaller companies compete for customers directly with larger retailers. Phase two was the liberalisation of the retail gas market in April 2017, and Phase three was the unbundling of transmission, distribution, and generation in 2020 (Chapman & Itaoka, 2018). Gas market reform followed with similar objectives – securing additional supply, expanding consumer choice, and decreasing gas prices, which have been among the highest in the OECD (IEA, 2016b). For over 50 years, regulations in the gas sector allowed gas utilities to operate monopolistically within their regions and service. In 1995, reforms opened competition in the natural gas sector for large customers, leading to deregulation of most of the gas retail market, dominated by Tokyo Gas, Osaka Gas, and Toho Gas, representing more than 70% of the market in 2015. In 2017, there were eight gas supply regions in Japan, with 26 public and 181 private gas utilities (EIA, 2017; IEA, 2016b; Kucharski & Unesaki, 2018).

The Gas Business Act was amended in 2017 to liberalise the gas market, initially for residential and small-business customers, by opening up the ¥2.4 trillion retail market to competition, gradually leading to a fully open market that included third-party access to LNG terminals (EIA, 2017; IEA, 2016b, 2021a). The new measures led to an increase in the share of new entrants and the number of retail business operators and customers switching. Regional monopolies began to break down, a trend strengthened further by government plans to fully separate the pipeline business division of large gas suppliers by 2022 to increase market competition (Kucharski & Unesaki, 2018). Unbundling represents the final phase of the natural gas market's reform, designed to increase third-party access to LNG terminals and increase competition in the natural gas sector.

Cost of fossil fuel imports and electricity prices

The cost of fuel imports has always placed a heavy burden on Japan's finances. In the three years following Fukushima, Japan spent approximately US$30 billion per year on fuel imports, with US$10 billion extra to meet the shortfall left by the sudden shutdown of nuclear generation (EIA, 2017). This increased dependence on imported fossil fuels led to the country's first trade deficit since the second oil shock in 1979, in stark contrast to the US$65 billion trade surplus in 2010 (Colombo et al., 2016; EIA, 2017). The subsequent electricity price increase led to greater public debt and revenue losses for utility companies (EIA, 2017). The average power price rose by approximately 20% for households and 30% for industry between 2011 and 2013 (Colombo et al., 2016), and electricity prices as a whole increased by 23% for homes and 27% for industry between 2010 and 2018 (Agency for Natural Resources and Energy Japan, 2019; METI, 2019).

A continued closure of nuclear facilities or prolonged ramp-up will affect the economy nationally and locally. In 2015, the nuclear industry employed 10,500 people – 3.1% of all employees in the electricity, gas, heat supply, and water sector (McLellan et al., 2013). METI (2015) concluded that the replacement cost of substituting nuclear energy with renewables would be approximately double the cost of LNG and nearly seven times that of coal-fired power generation. Nuclear power would reduce the energy system's levelised costs of electricity as nuclear capacity increases, even factoring in restoration and compensation costs of US$257 billion (Hong et al., 2013).

While there is scope for improvement to reduce renewable energy costs, renewables are not the only answer, and renewable energy is considered inadequate to supply the necessary energy for Japanese society, mainly because of an inconsistent FIT system and complicated regulations for solar and wind. At the same time, Japanese stakeholders question the country's readiness to move away from coal considering METI's recent coal phase-out announcement and increased climate-related concerns. Japan's decades-long policy of a balanced energy-mix approach combined with coal's overall cost competitiveness are factors that could make it more challenging to reduce coal consumption by as much as some analysts predict.

Impact on LNG

An LNG demand shock occurred following the country's nuclear powerplant shutdown when the north-east Asia LNG spot price reached over US$19.7/MMbtu in February 2013, up from US$7/MMbtu in 2010. Post-Fukushima, the overall cost of LNG imports doubled from US$35 billion in 2010 to US$75 billion in 2013 (Colombo et al., 2016; Len & Nian, 2014). Additional LNG was imported by utilities at high crude Japan Crude Cocktail-indexed prices, which directly and adversely affected the economy at a time of ongoing energy sector reforms (Carriere, 2018). Using LNG solely to replace nuclear gaps is unlikely and unrealistic owing to economic, structural, and security of supply concerns. Long-term exposure to volatility and rising LNG import costs would also affect the manufacturing sector and competitiveness of its exports.

The higher cost to the economy increased scrutiny of the long-standing LNG pricing structure. The Fukushima disaster and its aftermath "proved to be the tipping point that accelerated reforms at the policy level" for LNG contracting (Carriere, 2018, p. 140). This led to the liberalisation of the LNG market with a more market-based structure to optimise prices and quantities to directly benefit Japanese utilities. In the past, contracts had destination restrictions that limited Japanese buyers from reselling cargoes or taking deliveries of cargoes outside their original port without prior approvals. These clauses were included by sellers to limit buyers' ability to compete in other markets (Colombo et al., 2016). For decades, Japanese buyers have sought to increase flexibility in LNG contracts, initially focusing on delinking LNG prices from crude oil-based formulas in line with global trends and to cater for domestic electricity and gas reforms. Recently, some companies have developed joint procurement arrangements with other importers to improve procurement terms and increase flexibility. The JERA join venture in 2015 represented 40% of LNG imports and focused on centralising purchasing activities and creating market efficiencies (Colombo et al., 2016; EIA, 2017). These moves signalled that importers would seek to reduce their dependence on long-term contracts and increase their presence in the short-term market. Japanese buyers pushed back on destination restrictions in contracts as LNG supplies increased, and arbitrage opportunities were left to traders and sellers (Carriere, 2018). The China National

Offshore Oil Corporation (CNOOC) and the Korean Gas Company (KOGAS) signed a memorandum of understanding on LNG purchasing cooperation to leverage their collective efforts and market power. Stern (2016) predicted that tensions between LNG sellers and Japanese buyers would continue as pricing transitions from the traditional crude indexed system to a hub-based liquid market system.

Impact of competition on the energy transition

Reduced revenues linked to market liberalisation could negatively affect companies' ability to invest in infrastructure, consequently slowing the energy transition. One energy expert noted that post-Fukushima, METI has more leverage to liberalise, which may negatively affect incumbents, and that the dynamic could swing the balance of power from power companies to new entrants (Al-Kuwari, 2023). Similar observations highlight the belief that additional participants will disrupt a reliable system, built over decades. Despite plans to upgrade infrastructure to support deregulation, the country lacks a well-developed cross-country gas grid. A national grid is the most efficient and important tool to enhance competition, and some stakeholders consider it should be a government priority (Al-Kuwari, 2023).

Despite increased public concern about nuclear, and environmental awareness, economics is still the main consideration in shaping the energy debate. Many Japanese people are concerned about climate issues, but that is not their priority – economics remain central to future planning. Deregulation can lower prices globally, and consequently for consumers, making it a more competitive long-term option. Without continued government support in the shape of a carbon tax (or similar measures to support the energy transition), coal will remain an attractive option for energy importers.

Energy-transition dynamics

MLP analysis

Landscape analysis – social trends affecting Japan's energy system

Path dependence created by the "iron triangle" dynamic (combined with homogeneity in Japanese society and partially developed through national policy), the long reign of the LDP, and consistent communication in shaping public understanding of the country's policy goals all have contributed to a stable energy system that has only undergone gradual and planned changes over previous decades. The Kyoto Protocol and the Fukushima accident were two major exogenous factors affecting Japan's energy landscape (Chapman & Itaoka, 2018). Following Fukushima, METI responded to calls for increased transparency and dialogue in the energy policy debate, highlighting ways in which government could focus more on societal engagement and acknowledging the need for improvements in information provision and encouraging dialogue (METI, 2018).

Despite the general understanding of the importance of economic considerations, increased climate awareness supports the idea of Japan as a global citizen willing to sacrifice for nature and the greater good. This movement also coincided with increased public engagement and activism in society. While activism is not the cultural norm in Japan, it is evolving and could impact energy policy further. The media also amplify these voices, moving away from the Government-aligned narratives of the past. A Japanese policy expert concluded that mass media's increased coverage of natural disasters, especially typhoons, is a key driver of the public's increased interest in sustainability (Al-Kuwari, 2023). A 2015 survey showed that 34% of Japanese people believe rich countries should do more about climate change, while 58% say developing countries should do just as much as wealthy nations (Stokes et al., 2015).

Increased environmental awareness, the country's low-carbon ambitions, and the anti-nuclear movement are contradictory dynamics, limiting the impact of these landscape pressures. The anti-nuclear movement was not able to sustain the momentum it gained after the Fukushima accident, partly owing to Japanese society's understanding of the wider role that nuclear power must have in the short to medium term for Japan to achieve its climate objectives. Coal-fired power generation is the most economic and readily available alternative, while the pace of growth of renewables is limited. These factors could slow down Japan's ability to ramp up nuclear power and achieve its GHG emissions targets, while rendering it unable to shut down its nuclear power fleet completely to depend on coal, oil, and natural gas in the short to medium term.

Innovation and the R&D ecosystem

Japan has a rich history of R&D investment. R&D and innovation investment are key factors in successfully delivering the 2018 SEP, which concluded that a 2030 energy mix can be met using existing technologies, but to achieve its 2050 energy-mix targets, new technologies will be required (METI, 2018). Japan has the economic imperative and advanced industrial base, and a need to support large densely populated cities, to continue to lead on energy innovation. Historically, the majority of energy R&D was directed towards nuclear energy. Hydrogen, fuel cells, and solar PV are major new areas of focus. A significant long-term challenge is the lack of growth in R&D investment from the private sector (Tagami, 2020). In the energy-transition context, METI recognised that a more dynamic domestic environment that encourages investment in higher-risk industries, such as batteries, hydrogen development, and next-generation renewables, must be established to support Japanese companies if global competitiveness is to be maintained. This is reflected to some degree in the shift of R&D investment categories (METI, 2018).

Hydrogen

Hydrogen is a major focus in the country's future energy plans. The updated Basic Hydrogen Strategy released in 2023 (METI, 2023) confirms Japan's plan to be the world's first hydrogen society. The plan aims to accelerate investment in Japan's supply chain, encouraging US$113 billion public and private investment over the next 15 years (EIA, 2023; IEA, 2023; MFAT, 2023). The current phase is focused on increasing hydrogen production and use, the development of fuel cell electric vehicles and residential fuel cells, and the installation of electrolysers. By 2030, the focus is due to shift to the development and integration of hydrogen supply chains into the energy system. The final phase will centre around the development of a carbon-free hydrogen supply chain by 2040 (Dodds et al., 2015; IEA, 2016b; METI, 2023; MFAT, 2023). Overall, over US$1.5 billion has been dedicated to hydrogen R&D and subsidies to reduce the cost of zero-emission hydrogen for imports, develop import and domestic hydrogen infrastructure, and increase hydrogen use in key sectors including power generation, residential combined heat and power, and transport (Iida & Sakata, 2019; Nagashima, 2018). Hydrogen's attractiveness to Japanese policymakers is partly due to its versatility. The power sector is the key to long-term hydrogen demand in Japan and will account for nearly two-thirds of all demand by 2050 (Nagashima, 2018). Coal and natural gas plants can also be retrofitted to burn ammonia alongside fossil fuels for power generation to reduce CO_2 emissions and extend the life of older plants in a cost-effective manner. Ammonia could play a transition role as a carrier, especially for utilities looking to use carbon-free ammonia as a mixed fuel in their coal plants. Japan could follow the north-western European model of introducing hydrogen to industrial-cluster applications, such as around Tokyo and Yokohama (Lambert, 2020). Subject to policy and financial challenges, hydrogen could account for nearly one-third of Japan's total primary energy supply, and up to two-thirds of power generation by 2050 while playing a positive role in meeting its 80% target of reducing GHG emissions (Iida & Sakata, 2019; Nagashima, 2018).

Fuel source, country of origin, and method of production are the key variables in determining supply costs. Hydrogen's current retail price is approximately ¥100 per normal cubic metre $\left(Nm^3\right)$ or $\$0.90 / Nm^3$, with a target to reduce this to $¥30 / Nm^3$ by 2030 and $¥20 / Nm^3$ in the long term (METI, 2023; MFAT, 2023; Nagashima, 2018). Such major cost reduction targets require strong commitment in R&D and consistent scale-up of the hydrogen supply network, domestically and internationally. Imports will be needed given the limited projected land available to generate the amount of renewable energy required. Japan is one of 39 countries that is unable to generate 50% of electricity and 100% of its hydrogen from renewable energy sources by 2050 (Bloomberg, 2020). Imported green hydrogen could compete with the cost of fossil fuel-produced blue hydrogen where CCS is expensive in those countries with gas and coal. The landed cost of renewables-produced hydrogen from north-west Australia shipped to Japan will be US$2.81/kg by 2050, higher than the US$1.61/kg it is forecast to cost to produce it domestically (Bloomberg, 2020).

Carbon capture and storage

CCS is another key technology to help Japan achieve an 80% reduction in GHG emissions by 2050 (Iida & Sakata, 2019). It is also important in Japan's hydrogen plans, with the updated Basic Hydrogen Strategy identifying CCS as essential for producing low-carbon hydrogen (METI, 2023). The availability of CCS is critical for increasing the viability of using blue hydrogen via imported LNG, which could also allow for increased utilisation of the natural gas network. Japan lies in a geologically restricted area for large-scale storage potential. In 2000, a METI-led R&D programme identified an estimated 146 Gt of CO_2 storage capacity in onshore and offshore saline aquifers (IEA, 2016b). However, further assessment conducted by METI in 2023 increased this potential storage capacity to 240 Gt (Otsuki et al., 2024). In 2014, METI and the MoE launched a geological survey to identify three CO_2 storage sites around Japan (IEA, 2016b), with the first demonstration project completed in 2019 in Tomakomai (IEA, 2021a). The development of large-scale CCS is expensive, which leads to the preferred approach of using overseas CCS facilities (Iida & Sakata, 2019). Future management pathways include using post-combustion CO_2 for commercial or industrial products, and transporting CO_2 to neighbouring countries with better geological conditions.

Hydrogen can be viewed as the ultimate fuel and a realistic solution (for some, the only solution) that can play a positive role in meeting several objectives, mainly energy security by coexisting with, and eventually replacing, fossil fuels. Industry participants were sceptical, however, about the timeline in which hydrogen can play an important role in the energy system. One energy executive concluded that "even in 2050, I'm not so sure hydrogen can satisfy the need for societal activities" (Al-Kuwari, 2023). Despite Japan's focused long-term plans, hydrogen is not ready for commercialisation in the foreseeable future. A long-term hydrogen economy objective is seen as supportive to sustained (and even increased) LNG imports into Japan. Achieving hydrogen objectives without LNG in the short to medium term will be challenging. We need LNG to realise a hydrogen society. Hydrogen is expected to be a factor in the medium term. There is also an expectation that experience and know-how acquired from the development of the LNG industry can be applied to the creation of complex contractual relationships and supply chains for hydrogen.

Regime status summary

Japan's most recent renewable energy targets aim to increase the share of renewable energy generation to 36–38% by 2030 (EIA, 2024), increasing the level of ambition set out in its Long Term Energy Supply Outlook (METI, 2015); however, this remains below the threshold of the twin objectives of a new low-carbon regime and complete energy transition (Chapman & Itaoka, 2018). Niche innovations in renewables and hydrogen technologies will increase their role over time and move the energy system towards a reconfigured pathway to meet climate and energy goals. Satisfaction with Japan's energy policy direction could discourage significant energy system structural changes, giving natural gas an opportunity to remain in the short to medium term. LNG may still benefit from a positive image and the fact that the value of natural gas is widely shared among Japanese people; the perception of gas in Japan is very different to that in Europe. MLP analysis for Japan reveals the range of cultural, institutional, and environmental factors that impact the country's energy transition.

Transition context

Adapting Smith et al.'s (2005) governance of sustainability transition framework, this section presents an analysis of Japan's current system's readiness and future trajectory within the energy-transition context. The two main criteria for evaluation are Japan's energy-transition ambition and its adaptive capacity.

Energy-transition ambition

Japan's energy-transition ambition is perceived differently by stakeholders. A common theme is the concept of "two Japans" coexisting from an energy policy standpoint, with one representing innovation, cleaner technologies, and climate leadership, with hydrogen as the central focus, and the other representing a more efficient yet still fossil-fuel-dominated world. This duality is reflected in Japan's energy policies, which are still mixed in terms of this focus on cleaner technologies and has a foot in the old fossil-fuel-dominated world.

A pragmatic approach is recommended by some in Japan, taking into consideration that the energy plan must consider the current and potential future energy mix. The structure of the energy plan adds unnecessary planning complexity for companies because, from the private sector's point of view, nobody expects that nuclear power can be developed, and government plans are not private-sector friendly. An unclear future and multiple tensions place policymakers at a crossroads regarding the direction to take. Key uncertainties in energy supply are linked primarily to nuclear capacity ramp-up and the impact of deregulation. Management of variables is the biggest uncertainty in energy supply and demand.

Japan's depth of expertise and experience in the energy industry are national resources and assets that can be used to support a transition. The country is a great importer of coal, LNG, oil and nuclear fuel, and has very experienced procurement and marketing teams. Japanese companies' LNG expertise was developed with the domestic market in mind. However, they seek ways to expand natural gas's effective use in emerging economies to achieve a low-carbon society as a start and a decarbonised society as the final goal. Japan has also developed a strong energy infrastructure, identified as a key resource by several interviewees in Al-Kuwari (2023). Significant LNG regasification infrastructure and capacity, underpinned by over 30 LNG import terminals, represent significant assets. Market liberalisation and third-party access could unlock the potential of excess unutilised capacity and increase market participation. Although there is significant debate around the future of nuclear power, Japan's nuclear plant infrastructure is a resource and regardless of the current political climate, the country has a ready nuclear fleet that is low-carbon intensive. However, increased efforts by METI to liberalise the gas and power markets and shift the balance of power from established companies to new entrants could adversely affect infrastructure investment, ultimately slowing down energy-transition efforts.

Adaptive capacity – path dependency and the pace of change

Japan has a historically predictable energy strategy, subject to significant change in times of major shocks (Valentine et al., 2011). Its institutional culture of path dependence, however, supports incremental change over fundamental change, as major policy shifts imply an admission that previous senior policymakers made mistakes (Trencher et al., 2020). Structural change is difficult to enforce, particularly in the context of energy policy because path dependency is supported by the formation of interest groups and the size and complexity of energy systems (Okura et al., 2016). In the energy-transition context, stability is a major potential barrier to change. Industry participants interviewed consistently cautioned that Japan could not and should not transition fast (Al-Kuwari, 2023). The Japanese system requires time to develop, and the actors involved are not accustomed to sudden changes; a balance between existing and new systems is needed.

Al-Kuwari's (2023) interviews revealed unique characteristics linked to an ageing and declining population, a developed energy system, limited renewables potential, and complex energy priorities – all reasons why Japan should be viewed differently to emerging and European countries. One energy expert concluded: "If you compare the energy transitions in Japan with the European [one], it may be a little bit slow. But compared to the other countries in Asia, I think we have done a fair amount of transition stuff already" (Al-Kuwari, 2023). Another senior energy executive explained that Japan must create its own pathway: "We are aiming at an energy transition, but in a different way". The belief that Japan should not transition too fast could prove favourable to LNG in the short to medium term because LNG is seen as the next best option when compared with coal and nuclear. Longer term, the country's ambitious hydrogen plans could be at odds with Japan's cautious approach to the energy transition. Table 4.1 summarises Japan's adaptive resources and selection pressures.

Using Smith et al.'s (2005) transition context model, Kucharski and Unesaki (2018) classified Japan's energy transition as a "purposive transition", mainly owing to high levels of coordination in the form of policy direction and regulation and with dedicated new financial incentives and resources designed to lead to positive outcomes in the wider economy (Berkhout et al., 2013; Kucharski & Unesaki, 2018; Smith et al., 2005). The Japanese approach to the energy transition is undoubtedly geared towards a low-carbon transition, despite the focus on a "3E+S" approach that requires multiple competing objectives to be addressed concurrently. Japan has made significant GHG emissions-reduction commitments and has continuously signalled its desire to move towards a low-carbon economy.

Despite significant financial resources, Japan's internal resources profile is challenging. The country may be unable to use its substantial low-carbon nuclear-powered plant infrastructure and has no existing domestic resources with which to replace it. Furthermore, while Japanese companies are leaders in traditional energy, they have not redirected their focus to the energy transition like their European counterparts have done. Despite a potential loosening of the

Table 4.1 Summary of Japan's adaptive resources and selection pressures (developed by Al-Kuwari, 2020)

Adaptive capacity	
Creation of new knowledge	• Highly educated and skilled workforce • Sophisticated energy importers and traders • Hydrogen economy – ability to leverage LNG expertise
Resource supply	• Potential for increased low-carbon electricity, primarily offshore and onshore wind • Hydrogen with CCS (via imports) and electrolysis
Infrastructure	• 30+ LNG terminals and associated infrastructure with spare capacity • Low-carbon nuclear power plant infrastructure (a sunk cost to be leveraged)
Creation of positive external economies	• Diversification into a low-carbon system to create opportunities downstream in energy efficiency and transport • Long-term reduction of energy dependence lowers import bill and supports investment in domestic opportunities for low-carbon development
Market formation	• Next-generation expertise in energy efficiency, solar technology, battery storage • International hydrogen supply chain • CCS and hydrogen production – creation of new demand • Offshore wind – expansion of markets
Innovation policy interventions (i.e. R&D)	• Leadership in energy R&D and innovation • Basic hydrogen plan • 3E + S strategy
Selection pressures	
Social changes (landscape)	• Climate change – increased public/media awareness and pressure • Fukushima – anti-nuclear movement • International commitments – Kyoto Protocol, Paris Agreement
Economic (competition from other systems)	• Supply/demand imbalance – increased dependence on foreign supplies risks exposure to fossil fuel price fluctuations • Import costs
Creation of socio-technical reconfiguration within niches	• Industrial clusters – provide an opportunity for targeted policies to promote industrial development • Hydrogen economy – reconfiguration of Japan's entire energy system
Innovation that seeds a transformation	• Hydrogen and fuel-cell-related technologies could lead to breakthroughs in heating, industrial, and transportation uses

"iron triangle"-related closeness among key participants, and the threat of market deregulation for incumbents, the focus remains on the tools available and resisting rapid changes when possible. Based on this analysis, Japan is in a "purposive transition", with some elements pushing it towards "endogenous renewal".

Actors

Energy companies continue to dominate Japan's energy system. Both utilities and gas companies are powerful within the regime, although their power is decreasing as market deregulation opens to third parties and independent producers. Exporters also play a large role, and although LNG exporters are individually weak, their agency is collectively important. Politicians will set the tone for future ambitions and international agreements and can be influenced by the media and public opinion. The financial/investment community, academics, and NGOs have small but growing roles within the regime.

Summary of the T-E-T-I analysis

The Japan case highlights factors determining LNG's potential future pathways within its energy system in the energy transition (Figure 4.2). For policymakers and LNG importers, LNG provides a tool to mitigate the short- and medium-term impact of nuclear energy shortfall post-Fukushima and to support energy security overall. Longer term, the role of LNG will be determined by its competitiveness, and the industry's ability to reduce GHG emissions within the context of the country's decarbonisation ambitions.

Japan's slow consensus approach has affected its ability to adjust to serious supply shocks related to nuclear shortages. Similarly, the country did not exploit its deep technical and electronics expertise to fully develop its renewable energy industry. However, going forward, its ability to develop consensus between polity,

Resources

- No gas pipeline connections; developed LNG regasification infrastructure
- Lack of gas storage and cross-country pipeline systems
- Nuclear start-up and ramp-up is key in determining LNG's future role in power generation
- Coal's future role unclear; coal-to-gas switching provides emission savings in the short term
- Renewable growth will encounter geographical and infrastructure constraints
- LNG is a potential bridge for hydrogen economy

Market

- Large utilities and gas companies' influence reduced post-Fukushima and with impact of regulation
- Funding for coal development continues but is reduced
- Gas still expensive in relation to OECD; EU market regulation must be supported with infrastructure development
- Market liberalisation and gas reforms should lead to linking import prices to global LNG prices
- JKM increasing in relevance as a price market but not a pricing hub

Energy-transition Dynamics

- Increased awareness and interest in climate change
- Anti-nuclear movement supports environmental awareness
- 3E+S energy policy places equal importance on energy security; competitiveness could slow down pace
- LNG industry has expertise and resources, including infrastructure, to potentially adapt
- Potential disruptive technology niches with hydrogen know-how and investment
- Commitment to R&D and innovation funding is consistent

Power

- Politicians dealing with competing pressures
- Influence of bureaucracy, mainly METI and more increasingly MoE
- 'Iron Triangle' dynamic weakening; less aligned interests could hamper decision-making and investment in infrastructure
- Increased third parties and independent power producers could shift balance of power away from incumbents
- Role of media increasing; influencing end-user decisions and elevating public debate on energy policy

Figure 4.2 Japan case study following T-E-T-I framework analysis

bureaucracy, and business could provide a clear and less volatile path for development based on the country's priorities. Energy security remains the main priority for its energy policy – re-emphasised during the European energy crisis. The consensus around ammonia as a path for hydrogen development via imports allows for certainty in investment and planning, and aligns with the importance of nuclear as part of the energy mix. Within this context, LNG and LNG infrastructure could support the transition towards the hydrogen economy, while providing flexibility to support seasonality.

References

ACIL Allen Consulting for Arena. (2018). *Opportunities for Australia from hydrogen exports*. https://arena.gov.au/assets/2018/08/opportunities-for-australia-from-hydrogen-exports.pdf

Agency for Natural Resources and Energy Japan. (2019). *Energy white paper 2019 presentation*. www.meti.go.jp/english/report/downloadfiles/2014_outline.pdf

Al-Kuwari, O. (2020, May 18). *Our new energy transformation can see the Middle East thrive – if it invests in RDI and education*. Qatar Foundation Op-Ed. www.qf.org.qa/stories/op-ed-our-new-energy-transformation-can-see-the-middle-east-thrive-if-it

Al-Kuwari, O. (2023). *The future of liquified natural gas (LNG) in the energy transition: Options and implications for the LNG industry in a decarbonising world* [Unpublished doctoral dissertation]. University College London. https://discovery.ucl.ac.uk/id/eprint/10173650/

Arima, J., Hombu, K., & Tachibana, Y. (2015). *Is Japan's INDC "underwhelming"?* International Environment and Economy Institute.

Aruga, K., Islam, M. M., & Jannat, A. (2024). Assessing the CO2 emissions and energy source consumption nexus in Japan. *Sustainability*, *16*(13), 5742. https://doi.org/10.3390/su16135742

Berkhout, F., Smith, A., & Stirling, A. (2013). Socio-technological regimes and transition contexts. In B. Elzen, F. Geels, & K. Green (Eds.), *System innovation and the transition to sustainability* (pp. 48–75). Elgar Online. https://doi.org/10.4337/9781845423421.00013

BfS (Bundesamt für Strahlenschutz). (2022, February 15). *Health consequences of the accident of Fukushima*. www.bfs.de/EN/topics/ion/accident-management/emergency/fukushima/health-effects.html

BfS (Bundesamt für Strahlenschutz). (2024, February 21). *Environmental impact of the Fukushima accident: Radiological situation in Japan*. www.bfs.de/EN/topics/ion/accident-management/emergency/fukushima/environmental-consequences.html

Bloomberg. (2020). *Hydrogen economy outlook – key messages*. BloombergNEF. https://data.bloomberglp.com/professional/sites/24/BNEF-Hydrogen-Economy-Outlook-Key-Messages-30-Mar-2020.pdf

Boersma, T., & Jordaan, S. M. (2017). Whatever happened to the golden age of natural gas? *Energy Transitions*, *1*(2), 1–8. https://doi.org/10.1007/s41825-017-0005-4

Botzen, W., Gowdy, J., & Van den Bergh, J. (2008). Cumulative CO2 emissions: Shifting international responsibilities for climate debt. *Climate Policy*, *8*(6), 569–576. https://doi.org/10.3763/cpol.2008.0539

BP (British Petroleum). (2021). *BP statistical review of world energy 2021*. www.bp.com/content/dam/bp/business-sites/en/global/corporate/pdfs/energy-economics/statistical-review/bp-stats-review-2021-full-report.pdf

CarbonBrief. (2018, June 25). *The carbon brief profile: Japan*. www.carbonbrief.org/carbon-brief-profile-japan/

Carriere, C. (2018). The effects of Japan's push for greater LNG market flexibility on LNG pricing and destination restrictions. *Journal of World Energy Law & Business*, *11*(2), 136–144. https://doi.org/10.1093/jwelb/jwy002
Chapman, A. J., & Itaoka, K. (2018). Energy transition to a future low-carbon energy society in Japan's liberalizing electricity market: Precedents, policies and factors of successful transition. *Renewable and Sustainable Energy Reviews*, *81*, 2019–2027. https://doi.org/10.1016/j.rser.2017.06.011
Cherp, A., Vinichenko, V., Jewell, J., Brutschin, E., & Sovacool, B. (2018). Integrating techno-economic, socio-technical and political perspectives on national energy transitions: A meta-theoretical framework. *Energy Research and Social Science*, *37*, 175–190. https://doi.org/10.1016/j.erss.2017.09.015
Colombo, S., El Harrak, M., & Sartori, N. (Eds.). (2016). The future of natural gas: Markets and geopolitics. *European Energy Review*. https://doi.org/10.1016/0040-1625(87)90047-3
Corbeau, A., & Ledesma, D. (Eds.). (2016). *LNG markets in transition: The great reconfiguration*. Oxford University Press for the Oxford Institute for Energy Studies and the King Abdullah Petroleum Studies and Research Center.
Dodds, P. E., Staffell, I., Hawkes, A. D., Li, F., Grünewald, P., McDowall, W., & Ekins, P. (2015). Hydrogen and fuel cell technologies for heating: A review. *International Journal of Hydrogen Energy*, *40*(5), 2065–2083. https://doi.org/10.1016/j.ijhydene.2014.11.059
Efird, B., Al Muhanna, S., Al-Mubarak, I., Turkistani, S., & Al-Ghamdi, F. (2018). *The policymaking process to restart Japanese nuclear power plants*. KAPSARC. https://doi.org/10.30573/KS–2018-DP47
EIA (United States Energy Information Administration). (2017). *Country analysis brief: Japan*. www.iberglobal.com/files/2017/japon_eia.pdf
EIA. (2023). *Country analysis brief: Japan*. www.eia.gov/international/content/analysis/countries_long/Japan/japan.pdf
EIA. (2024, May 2). *Japan's energy policies aim for increased zero-carbon electricity generation*. www.eia.gov/todayinenergy/detail.php?id=61944#:~:text=Policies%20target%20an%20increase%20in,%25%E2%80%9322%25%20by%202030
Hayashi, M., & Hughes, L. (2013). The Fukushima nuclear accident and its effect on global energy security. *Energy Policy*, *59*, 102–111. https://doi.org/10.1016/j.enpol.2012.11.046
Hong, S., Bradshaw, C. J., & Brook, B. W. (2013). Evaluating options for the future energy mix of Japan after the Fukushima nuclear crisis. *Energy Policy*, *56*, 418–424. https://doi.org/10.1016/j.enpol.2013.01.002
Hugo, I., & Kusumawardhana, I. (2023). Way of the samurai: Understanding Japan's energy security strategy post Fukushima disaster. *Journal of Social Political Sciences*, *4*(1), 81–97. https://doi.org/10.52166/jsps.v4i1.156
IEA (International Energy Agency). (2016a). *Energy and air pollution – world energy outlook: Special report 266*. IEA Publications. https://iea.blob.core.windows.net/assets/6b75c4ae-e633-4fa0-9569-b28e226e6103/WorldEnergyOutlookSpecialReport2016EnergyandAirPollution.pdf
IEA (International Energy Agency). (2016b). *IEA Japan energy policies 2016*. IEA Publications. https://iea.blob.core.windows.net/assets/0544f290-7b29-4733-a6c4-fb18d33b0ac4/EnergyPoliciesofIEACountriesJapan2016.pdf
IEA (International Energy Agency). (2018). *Outlook for producer economies*. IEA Publications. www.iea.org/reports/outlook-for-producer-economies
IEA (International Energy Agency). (2019). *World energy investment 2019*. IEA Publications. www.iea.org/reports/world-energy-investment-2019
IEA (International Energy Agency). (2021a). *Japan 2021 energy policy review*. IEA Publications. https://iea.blob.core.windows.net/assets/3470b395-cfdd-44a9-9184-0537cf069c3d/Japan2021_EnergyPolicyReview.pdf

IEA (International Energy Agency). (2021b). *World energy outlook 2021*. IEA Publications. www.iea.org/reports/world-energy-outlook-2021

IEA (International Energy Agency). (2023). *World energy outlook 2023*. IEA Publications. https://iea.blob.core.windows.net/assets/86ede39e-4436-42d7-ba2a-edf61467e070/WorldEnergyOutlook2023.pdf

IEA (International Energy Agency). (n.d.). *Japan*. Retrieved July 9, 2024, from www.iea.org/countries/japan

IEEFA (Institute for Energy Economics and Financial Analysis). (2024). *Global LNG outlook 2024–2028*. https://ieefa.org/sites/default/files/2024-04/Global%20LNG%20Outlook%202024-2028_April%202024%20%28Final%29.pdf

IEEJ (Institute of Energy Economics Japan). (2021). *IEEJ outlook 2022: Challenges toward carbon neutrality*. https://eneken.ieej.or.jp/en/whatsnew/439.html

Iida, S., & Sakata, K. (2019). Hydrogen technologies and developments in Japan. *Clean Energy*, *3*(2), 105–113. https://doi.org/10.1093/ce/zkz003

IRENA (International Renewable Energy Agency). (2024, April 2). *Country rankings*. www.irena.org/Data/View-data-by-topic/Capacity-and-Generation/Country-Rankings

Ishimoto, Y., Kurosawa, A., Sasakura, M., & Sakata, K. (2017). Significance of CO2-free hydrogen globally and for Japan using a long-term global energy system analysis. *International Journal of Hydrogen Energy*, *42*(19), 13357–13367. https://doi.org/10.1016/j.ijhydene.2017.02.058

Jensterle, M., Narita, J., Piria, R., Samadi, S., Prantner, M., Crone, K., Siegemund, S., Kan, S., Matsumoto, T., Shibata, Y., & Thesen, J. (2019). The role of clean hydrogen in the future energy systems of Japan and Germany. *International Journal of Hydrogen Energy*, *44*(1), 559–571. https://d-nb.info/1198308249/34

Kucharski, J. B., & Unesaki, H. (2018). An institutional analysis of the Japanese energy transition. *Environmental Innovation and Societal Transitions*, *29*, 126–143. https://doi.org/10.1016/j.eist.2018.07.004

Lambert, M. (2020). *Hydrogen and decarbonisation of gas: False dawn or silver bullet?* Oxford Institute for Energy Studies. www.oxfordenergy.org/wpcms/wp-content/uploads/2020/03/Insight-66-Hydrogen-and-Decarbonisation-of-Gas.pdf

Len, C., & Nian, V. (2014). Nuclear versus natural gas: An assessment on the drivers influencing Japan's energy future. *Energy Procedia*, *61*, 194–197. https://doi.org/10.1016/j.egypro.2014.11.1060

McKinsey Global Institute. (2019). *Global energy perspective 2019: Reference case*. Energy Insights. www.mckinsey.com/~/media/mckinsey/industries/oil%20and%20gas/our%20insights/global%20energy%20perspective%202019/mckinsey-energy-insights-global-energy-perspective-2019_reference-case-summary.ashx

McLellan, B. C., Chapman, A. J., & Aoki, K. (2016). Geography, urbanization and lock-in – considerations for sustainable transitions to decentralized energy systems. *Journal of Cleaner Production*, *128*, 77–96. https://doi.org/10.1016/j.jclepro.2015.12.092

McLellan, B. C., Zhang, Q., Utama, N. A., Farzaneh, H., & Ishihara, K. N. (2013). Analysis of Japan's post-Fukushima energy strategy. *Energy Strategy Reviews*, *2*(2), 190–198. https://doi.org/10.1016/j.esr.2013.04.004

METI (Ministry of Economy Trade and Industry Japan). (2010). *2010 Japan's annual report on energy*. www.meti.go.jp/english/report/downloadfiles/2010_outline.pdf

METI. (2014). *METI 4th strategic energy plan*. www.enecho.meti.go.jp/category/others/basic_plan/pdf/20211022_01.pdf

METI. (2015). *Long-term energy supply and demand outlook*. www.meti.go.jp/english/press/2015/pdf/0716_01a.pdf

METI. (2018). *Japan's fifth strategic energy plan (provisional translation)*. www.enecho.meti.go.jp/en/category/others/basic_plan/5th/pdf/strategic_energy_plan.pdf

METI. (2019). *Japan's energy – 10 questions to understand the current energy situation*. www.enecho.meti.go.jp/en/category/brochures/

METI. (2022, October 26). *2021 – Understanding the current energy situation in Japan (part 1)*. www.enecho.meti.go.jp/en/category/special/article/detail_171.html

METI. (2023). *Basic hydrogen strategy*. www.meti.go.jp/shingikai/enecho/shoene_shinene/suiso_seisaku/pdf/20230606_5.pdf

MFAT (Ministry of Foreign Affairs and Trade). (2023). *Japan: Hydrogen strategy. Market intelligence report*. www.mfat.govt.nz/assets/Trade-General/Trade-Market-reports/Japan-hydrogen-strategy-November-2023.pdf

Ministry of Environment. (2017). *Outline of long-term low-carbon vision*. www.env.go.jp/press/103822/713.pdf

Moe, E. (2012). Vested interests, energy efficiency and renewables in Japan. *Energy Policy*, *40*(1), 260–273. https://doi.org/10.1016/j.enpol.2011.09.070

MOFA (Ministry of Foreign Affairs of Japan). (2021). *Climate change – Japan's initiative toward net-zero GHG emissions by 2050*. www.mofa.go.jp/files/100256809.pdf

Nagashima, M. (2018). *Japan's hydrogen strategy and its economic and geopolitical implications*. IFRI Centre for Energy. www.ifri.org/en/publications/etudes-de-lifri/japans-hydrogen-strategy-and-its-economic-and-geopolitical-implications

Nishimura, K. (2016). What was the "discrepancy" in Japan's energy policy after Fukushima nuclear crisis? *Economics of Energy and Environmental Policy*, *5*(1), 3–13. https://doi.org/10.5547/2160-5890.5.1.knis

Okura, S., Tkach-Kawasaki, L., Kobashi, Y., Hartwig, M., & Tsujinaka, Y. (2016). Analysis of the policy network for the "feed-in tariff law" in Japan: Evidence from the GEPON survey. *Journal of Contemporary Eastern Asia*, *15*(1), 41–63. http://eastasia.yu.ac.kr/documents/Okura_15_1.pdf

Oshiro, K., Kainuma, M., & Masui, T. (2016). Assessing decarbonization pathways and their implications for energy security policies in Japan. *Climate Policy*, *16*, S63–S77. https://doi.org/10.1080/14693062.2016.1155042

Otsuki, T., Shibata, Y., Matsuo, Y., Obane, H., & Morimoto, S. (2024). Role of carbon dioxide capture and storage in energy systems for net-zero emissions in Japan. *International Journal of Greenhouse Gas Control*, *132*, 104065. https://doi.org/10.1016/j.ijggc.2024.104065

Shibata, Y. (2016). *Is power to gas feasible in Japan ?* Institute for Energy Economics. https://eneken.ieej.or.jp/data/6549.pdf

Smith, A., Stirling, A., & Berkhout, F. (2005). The governance of sustainable socio-technical transitions. *Research Policy*, *34*(10), 1491–1510. https://doi.org/10.1016/j.respol.2005.07.005

Solomon, B. D., & Krishna, K. (2011). The coming sustainable energy transition: History, strategies, and outlook. *Energy Policy*, *39*(11), 7422–7431. https://doi.org/10.1016/j.enpol.2011.09.009

Stern, J. (2016). *The new Japanese LNG strategy: A major step towards hub-based gas pricing in Asia*. Oxford Institute of Energy Studies. www.oxfordenergy.org/wpcms/wp-content/uploads/2016/06/The-new-Japanese-LNG-strategy-a-major-step-towards-hub-based-gas-pricing-in-Asia.pdf

Stokes, B., Wike, R., & Carle, J. (2015, November 5). *Global concern about climate change*. Pew Research Center. www.pewresearch.org/global/2015/11/05/global-concern-about-climate-change-broad-support-for-limiting-emissions/

Tagami, T. (2020). *Outlook and challenges of climate change policies in 2020*. Institute of Energy Economics Japan. https://eneken.ieej.or.jp/data/8868.pdf

Taghizadeh-Hesary, F., Yoshino, N., & Rasoulinezhad, E. (2017). Impact of the Fukushima nuclear disaster on the oil-consuming sectors of Japan. *Journal of Comparative Asian Development*, *16*(2), 113–134. https://doi.org/10.1080/15339114.2017.1298457

Trencher, G., Rinscheid, A., Duygan, M., Truong, N., & Asuka, J. (2020). Revisiting carbon lock-in in energy systems: Explaining the perpetuation of coal power in Japan. *Energy Research and Social Science*, *69*, 101770. https://doi.org/10.1016/j.erss.2020.101770

Tusiani, M., & Shearer, G. (2016). *LNG: Fuel for a changing world: A nontechnical guide*. Pennwell Books.

Valentine, S. V., & Sovacool, B. K. (2010). The socio-political economy of nuclear power development in Japan and South Korea. *Energy Policy*, *38*(12), 7971–7979. https://doi.org/10.1016/j.enpol.2010.09.036

Valentine, S., Sovacool, B. K., & Matsuura, M. (2011). Empowered? Evaluating Japan's national energy strategy under the DPJ administration. *Energy Policy*, *39*(3), 1865–1876. https://doi.org/10.1016/j.enpol.2011.01.029

Vivoda, V. (2019). LNG import diversification and energy security in Asia. *Energy Policy*, *129*, 967–974. https://doi.org/10.1016/j.enpol.2019.01.073

World Nuclear Association. (2024, June 28). *Nuclear power in Japan*. https://world-nuclear.org/information-library/country-profiles/countries-g-n/japan-nuclear-power

Zhang, Q., Ishihara, K. N., & Tezuka, T. (2010). A feasibility study on a future zero-carbon electricity system based on 100% nuclear power in Japan. *Journal of Nuclear Science and Technology*, *47*(12), 1182–1192. https://doi.org/10.1080/18811248.2010.9720985

Zhu, D., Mortazavi, S. M., Maleki, A., Aslani, A., & Yousefi, H. (2020). Analysis of the robustness of energy supply in Japan: Role of renewable energy. *Energy Reports*, *6*, 378–391. https://doi.org/10.1016/j.egyr.2020.01.011

5 China

Introduction

China may overtake the United States as the main contributor of cumulative concentrations of CO_2 by the middle of this century (Botzen et al., 2008; Stevens, 2023). The transition towards a lower-carbon economy must address the role of coal and involve natural gas, nuclear, hydro, renewables, and energy efficiency (Corbeau & Ledesma, 2016). China's planned transition is part of a larger state and mission-driven "green development strategy" of effective resource management (Mazzucato, 2013), which aims to leverage the country's vast human and financial capital with a high degree of centralised institutional adaptability (Andrews-Speed, 2016). Regardless of China's short- and medium-term focus on air quality, its willingness to combat climate change is crucial because any long-term pathway to meeting the Paris Agreement's goals (UNFCCC, 2015) depends heavily on its stabilisation of CO_2 concentrations. LNG's medium- to long-term future in China will depend on its affordability versus all alternatives, and on China's long-term climate and decarbonisation strategy, which must also consider social and economic pressures. Again, the T-E-T-I framework is used to structure the analysis in this chapter.

Natural gas and LNG in China

History and context

China is the world's most populous country with 1,409 billion people in 2023 (Master, 2024). The majority of global industrial energy demand in the near future is predicted to come from Asian and African countries, with China and India expected to account for nearly half of the world's energy demand growth up to 2040 (BP, 2020; ExxonMobil, 2019). China is aware of the need for cleaner energy and emissions reductions in conjunction with this increased energy demand. In fact, climate change mitigation has developed into a transition opportunity for China's industry and economy to support the country's ambitions to be a global technology leader (Meidan, 2020). The country has supported the global development of renewable energy by growing its domestic market and exporting solar and wind energy components at scale (Wood Mackenzie, 2024). It also has a thriving

DOI: 10.4324/9781003545149-5

local market, leading to China's renewable energy manufacturers becoming global leaders in terms of market share (Kaaresvirta et al., 2021). Natural gas, the cleanest fossil fuel, has also been gaining prominence with demand growing in China at 13% per year, making it the third-largest natural gas consumer behind the United States and Russia, the fastest-growing gas market globally, and, given the current relatively low penetration, the market with the highest potential (EIA, 2020; Fulwood et al., 2020; IEA, n.d.).

China's energy mix

Demand for energy doubled in China between 1990 and 2010, making it the world's largest energy consumer and producer (EIA, 2020; IEA, n.d.). In per capita terms, China's economy has grown 24 times between 1978 and 2017 (Meidan, 2020), including a nearly fivefold growth between 1990 and 2010 (Gilbert & Sovacool, 2017; Meidan, 2020). China's economy grew by 6.1% in 2019 prior to the COVID-19 pandemic. This, however, was the lowest growth rate since 1990, and followed an average growth rate of 10% per year between 2000 and 2011 (EIA, 2020). Indeed, since 2014, the country has been experiencing slower growth, moving away from energy-intensive state manufacturing and infrastructure activities towards services and domestic consumption (Corbeau & Ledesma, 2016). The 2020 economic slowdown coincided with an attempt at financial regulatory reforms, a reduction in high levels of local government debt, and cuts to industry-generated air pollution levels. These factors increased the risk of rising unemployment levels (EIA, 2020) with energy- and labour-intensive industrial activities increasingly relocating from China to lower-cost economies (BP, 2020).

The total primary energy consumption in China has been steadily rising, from 43 exajoules (EJ) in 2000 to 183 EJ in 2020, a fourfold increase (EIA, 2023). The country is currently heavily dependent on fossil fuels, with coal supplying approximately 58% of the total energy consumed in 2021 (EIA, 2023). Natural gas, nuclear power, and renewable energy consumption have increased, while coal use has plateaued in recent years (EIA, 2023). In 2021, coal constituted 58% of the country's primary energy consumption, followed by oil (19%), natural gas (8%), and hydro-electric power (8%).

Role of coal

China is the world's largest coal importer with approximately 357 million short tonnes imported in 2021. However, imports decreased to 210 million short tonnes in 2022 owing to greater domestic production. In 2022, Indonesia supplied 89% of China's imported coal, providing low-quality coal that blends well with China's domestic coal. Australia, once the second-largest exporter of coal to China, did not supply any coal in 2022 owing to China's unofficial ban on coal from Australia (EIA, 2023). China is also the world's leading producer of coal, responsible for more than half of global production (IEA, 2023). China's domestic coal production has risen steadily since 2016, with an estimated 4.8 billion short tonnes produced in

2022 (EIA, 2023). The country's coal reserves are mainly distributed in Central and Western regions, leading to a higher proportion of coal use in those areas (Xu & Lin, 2019).The power sector consumed nearly 60% of China's coal in 2021, with the remainder used in industry (steel and cement production) and residential heating. Coal consumption grew in 2018 and 2019 after several years of decline, partly through electricity and industrial demand growth and, in particular, steel production (EIA, 2020). Coal's competitive price will likely lead to it retaining its leading role in power generation for the foreseeable future (Kaaresvirta et al., 2021).

Role of oil and liquids

Oil production at China's oldest and largest oil fields, including Daqing and Shengli, is declining, making it difficult to maintain or grow the country's oil fields (Downs, 2021). When commodity prices recovered in 2016, China's technically challenging fields became more profitable and led to a temporary increase in investment. However, any future investment is highly dependent on oil prices, as witnessed in 2020 following a drop in oil prices (EIA, 2020).

Role of renewables and nuclear

China is the world's largest producer of renewable energy, with wind and solar providing a combined installed capacity of 758 GW in 2022 (EIA, 2023). Energy security and climate change priorities have led to an increased deployment of renewable energy (Andrews-Speed, 2015). Increasing use of technology in manufacturing and other sectors is also driving greater electricity demand (EIA, 2020). In 2005, the Renewable Energy Law was implemented, setting targets for renewable energy capacity, obligations for grid operators, the legal basis for FITs, and establishing a research and innovation fund. In its "Medium to Long-Term Plan for Renewable Energy Development" the National Development and Reform Commission (NDRC) set out its priority to develop domestic manufacturing capacity to promote deployment by reducing costs through economies of scale (Andrews-Speed, 2015). Rural electrification supported renewable energy development, initially in the 1950s with small-scale hydropower, then in the 1990s with solar PV (Andrews-Speed, 2015). In 2013, new renewable power capacity in China exceeded the addition of new capacity from fossil fuels and nuclear energy for the first time (Isoaho et al., 2017). However, the relative role of renewables in electricity generation has not changed significantly, despite sizeable investment.

Insufficient grid capacity was a major barrier to renewable energy development in the mid-2010s. In 2015, 15% of wind and 9% of solar output were curtailed owing to grid constraints (Isoaho et al., 2017). Renewables are still heavily dependent on subsidies, which the Government started to cut late in the 2010s. Since 2011, subsidies have contributed to the growth of renewable projects but reduced competition in the sector (Kaaresvirta et al., 2021). Renewable energy has been promoted as an add-on to the existing system, and the associated failure to integrate it has led to its curtailment and inefficiencies (Zhongying & Sandholt,

2019). The system suffers from poor coordination between renewable installations, and grid construction is a barrier for development with most of the wind and solar potential located in the west, far from the regions where most is consumed (Kaaresvirta et al., 2021; Meidan, 2019). A lack of storage capacity and transmission grid restrictions have contributed to significant power wastage from renewable sources (Kaaresvirta et al., 2021). Most areas suitable for onshore wind power have been developed, with 80% of high-quality projects residing with state-owned entities (SOE) (Downs, 2021).

China has one-quarter of the world's hydropower production capacity, at 414 GW in 2022 (EIA, 2023). Hydropower is the highest contributor to renewable electricity production, although growth in hydroelectricity is forecast to slow given that most viable sites have already been developed (BP, 2020; Kaaresvirta et al., 2021).

Nuclear energy represents 5% of total power generation, increasing from 2% nearly a decade ago (BP, 2020; EIA, 2023; Kaaresvirta et al., 2021). China is one of few countries constructing new plants in the past 15 years, with 40 new reactors connected to the grid in the past decade. In 2022, 49 nuclear reactors supplied electricity to the grid at a capacity of 56 GW, up from 18 GW in 2010 (BP, 2020; EIA, 2020, 2023; Kaaresvirta et al., 2021). China actively promotes nuclear power as a clean and reliable source of electricity generation. It is viewed as a means to diversify clean electricity production as electricity demand and the economy continue to grow (EIA, 2020; Kaaresvirta et al., 2021; World Nuclear Association, 2024). Nuclear power also addresses geographical challenges linked to a mismatch in inland coal production with the majority of economic activity and electricity demand located in eastern coastal areas (Kaaresvirta et al., 2021). China's nuclear plans were adversely affected by the Fukushima disaster (see Chapter 4), which led to a reduction in targets and plans for new plants (World Nuclear Association, 2024). Nonetheless, the China Nuclear Energy Agency (CNEA) has forecast that nuclear power capacity will reach 240 GW by 2050, sufficient to meet 15% of the country's projected electricity demand (Kaaresvirta et al., 2021; World Nuclear Association, 2024). However, China's uranium production does not meet its current fuel demands. Therefore, uranium fuel is imported mainly from Kazakhstan and Namibia, and China has a supply agreement with Australia (Kaaresvirta et al., 2021). Hydrogen demand is due to increase in China at a similar rate (or faster) than that of developed economies (BP, 2020; IEA, 2022).

Carbon dioxide emissions

Since 2006, China has been the world's largest emitter of CO_2 and total GHGs, partly because of increased demand for electricity and increased industrialisation and urbanisation (Gilbert & Sovacool, 2017; IEA, 2023, n.d.; Meidan, 2020; Xu & Lin, 2019). The country's development and the subsequent increase in emissions widens the scale and scope of the mitigation challenge and the country's ability to meet its climate targets (Anderson & Bows, 2011). In addition to CO_2, as of 2015, China is the largest emitter of SO_2, NO_X, and $PM_{2.5}$. Manufacturing industries accounted for 44% of SO_2, almost 40% of NO_X and 35% of $PM_{2.5}$ emissions (IEA, 2016).

Natural gas dynamics and infrastructure

From 2000 to 2020, natural gas evolved from a minor domestically produced resource into a fundamental energy source for the Chinese energy market, and gas consumption grew at an average of 14% per year (Kaaresvirta et al., 2021; Miyamoto & Ishiguro, 2018). Low natural gas prices, increased use in the transport sector, and the expansion of natural gas infrastructure all were demand drivers for the fuel (EIA, 2023). Of the world's top-five gas consumers, China is the only country that has seen an annual growth rate above 10% between 1997 and 2015 (Jiang et al., 2020).

Production

China's natural gas production has been rising steadily with 12.8 TCF produced by national oil companies (NOCs) in 2021. Shale gas production reached 812 BCF in 2021 (Wei et al., 2022), and China's offshore natural gas production reached 335 BCF in 2019, mostly through growth from the South China Sea (EIA, 2020). In 2021, approximately 60% of China's natural gas consumption was met by domestic sources (EIA, 2023). Although conventional production could continue to grow in the short term, it will not be able to keep up with increasing demand (Sandalow et al., 2018). Financial incentives were introduced to promote domestic upstream development of unconventional natural gas, including the reduction of resource taxes and the extension of subsidies (EIA, 2020). Despite the potential for shale gas to improve energy security by providing up to 30 BCM per year (Sandalow et al., 2018), it is unlikely that extraction will increase dramatically given complex terrain, poor geology, high costs, long distances to market, and a SOE-dominated market landscape (Miyamoto & Ishiguro, 2018; Sandalow et al., 2018).

Sectors and regions

Industry accounts for approximately 56% of gas consumption, residential use 22%, and transportation 12% (IEA, n.d.; Kaaresvirta et al., 2021; Li, 2015). Industrial demand is rooted in the rise of more energy-intensive industries, while urbanisation and access to gas have increased demand from the domestic sector (Li, 2015). The residential sector has the highest price elasticity and is most sensitive to short- and long-term gas price changes (Lin & Li, 2020). In the power sector, natural gas's ability to provide cost-effective peak power, its low emissions compared with coal, and the numerous gas-fired power plants all drive demand growth. Furthermore, government schemes to promote natural gas use by heavy goods vehicles and LNG carriers support growth for this fuel (Li, 2015; Wang et al., 2020). Figure 5.1 depicts China's natural gas consumption.

High natural gas demand exists in regions close to production areas such as Sichuan and Chongqing, where long-distance pipelines are established and operational and there are well-developed LNG sales systems, mainly around coastal areas (Jiang et al., 2020). Demand is lower in China's Northeast and Northwest regions, where import capacity is relatively weak and slow economic growth leads

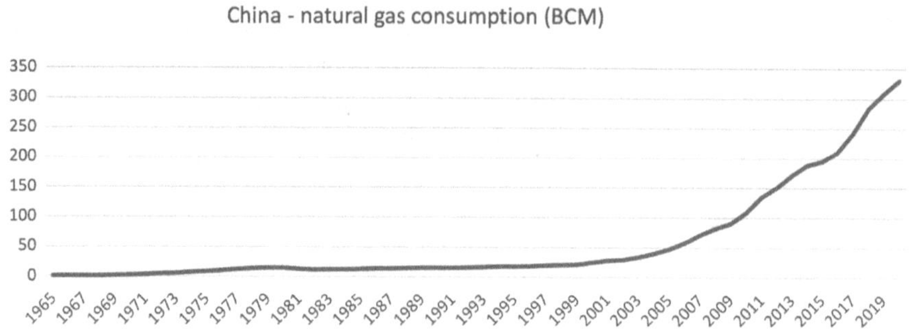

Figure 5.1 China's natural gas consumption trend (BP Statistical Review of World Energy, 2021)

to low or negative gas consumption growth (Jiang et al., 2020). The net exporting regions of Heilongjiang, Sichuan, Shaanxi, Qinghai, and Xinjiang are far from the coastal demand centres (Li, 2015). Shaanxi and Sichuan are two of the largest regions for both production and consumption, while Guangdong and Beijing are the largest gas-consuming regions, mostly for power and heat (Jiang et al., 2020; Li, 2015). Rapid development of the manufacturing sector and continuous urbanisation support high growth of natural gas consumption in the southeast region. In regions where economies are dominated by services, residential sector gas use is more prevalent (Jiang et al., 2020; Li, 2015).

Pipeline and LNG imports

China became a net importer of natural gas in 2007, and the world's largest importer in 2019, importing 5.6 TCF in 2023 (EIA, 2020; Energy Institute, 2024; Lin & Li, 2020). Imports are sourced primarily from Turkmenistan and Central Asian countries, and from Myanmar and Russia (EIA, 2020, 2023). The Central Asia–China Gas Pipeline comprises three pipelines (Lines A, B, and C) and is used for importing natural gas from Central Asia, flowing from Turkmenistan to China through Uzbekistan and Kazakhstan. The total planned capacity for this infrastructure system is 65 BCM per year (Miyamoto & Ishiguro, 2018). A fourth pipeline (Line D) was expected to add a further 1.1 TCF (31 BCM) capacity and increase the amount of natural gas from Turkmenistan by another 2.3 TCF per year (65 BCM). Dogged by delays, the pipeline is due to resume construction in 2024 and may be completed by the end of the 2020s (Liang, 2023). China extended its contract with Kazakhstan to 2023 and doubled the amount of imported gas to 350 BCF per year (EIA, 2020). In December 2019, China began importing from Russia via the Power of Siberia pipeline – a deal signed in 2014 to import 1.3 TCF per year (36 BCM) from Gazprom's East Siberian fields for 30 years (EIA, 2020). Power of Siberia 1 plans to supply 38 BCM per year for 30 years and will take several years to plateau following the start of operations (Miyamoto & Ishiguro, 2018). China plans to expand its side of the pipeline to deliver natural gas to

Beijing and other demand centres. The total length of natural gas pipelines was 122,000 km in 2023, with a further 66,000 km either in construction or proposed (Global Energy Monitor, 2023). The NDRC supports the expansion of the domestic natural gas pipeline network partly to ensure that gas can be exported to the northern parts of the country (Sandalow et al., 2018). There has been an increasing trend in China's LNG imports, which reached 84 BCM in 2022, with Australia providing the source for most of that gas (20%) followed by Qatar (15%), Malaysia (7%), and Russia (6%) (EIA, 2023).

Imported LNG demand is linked to increases in economic growth and energy consumption, growing dependence on imports, and coal-related air pollution (Gilbert & Sovacool, 2017). China imports from 20 LNG suppliers and five pipeline gas suppliers, with imports representing 42% of 2021 consumption (IEA, n.d.). By 2023, 27 LNG regasification terminals were in operation, with a capacity of 5.7 TCF (161 BCM). China is building additional terminals expected to add a further 5.5 TCF (155 BCM) of import capacity by 2026 (EIA, 2023). By the end of 2020, the country had 83.9 MTPA of regasification capacity, second only to Japan, and was forecast to add another 56 MTPA by 2024 (IGU, 2021).

Future growth in natural gas demand is expected to depend on China's environmental policies in the power sector and increased coal-to-gas switching for households and industry, the ability to build sufficient natural gas infrastructure, availability of affordable gas supply, and evolving environmental standards. Growth in LNG demand can be expected, as domestic supplies and pipeline imports will not meet demand and the lack of storage capacity will not support seasonal and short-term flexibility. With an accelerating low-carbon transition, dependence on coal will have to be reduced and options including renewables, nuclear, and hydropower need to be explored, although these options may potentially be limited by pricing structure (Corbeau & Ledesma, 2016; EIA, 2020; Meidan et al., 2021; Sandalow et al., 2018; Zou et al., 2018).

Natural gas forecasts

Energy demand growth in China is forecast to peak in the eary 2030s in all of BP's scenarios (BP, 2020). In its "rapid" and "net zero" scenarios, energy demand in China in 2050 is similar to 2018 levels owing to improvements in energy efficiency and a continued shift in the economy's structure. China remains the largest energy market in all scenarios, accounting for 20% of the world's energy demand (BP, 2020). In the China National Petroleum Company's (CNPC) 2050 report, gas demand growth was lowered from previous forecasts. However, demand is set to increase to 610 BCM in 2035 and 690 BCM in 2050, driven by the residential and power sectors. Gas import dependence is expected to reach 50% (Meidan, 2019). The evolution of import gaps and how LNG compares with alternative options will play a major role in regional and global LNG markets in future (Corbeau & Ledesma, 2016). Natural gas is also expected to play a strong bridging role in China when large-scale commercial CCS becomes available (McGlade & Ekins, 2014).

Insights from interviews

Demand for natural gas in China is unlikely to decrease in the coming decades. Participants in Al-Kuwari's (2023) study believe that new and additional applications for natural gas in the power and transport sectors are likely, and that the next 15–20 years could be "a golden period" for natural gas in China. Gas-fired power generation could grow because of increased capacity, gas pricing negotiating power, and the support offered by gas to support renewable growth through peak-load management assistance and a balancing of demand variations. Government policies favour natural gas over oil and coal, and it is unlikely from a technical perspective that gas consumers will be able to move fully to renewables in the short term. China's ambitious local and international sourcing of natural gas will support its growing economy. LNG demand seems set to increase in the next 20 years, with increasing numbers of buyers securing greater supplies and building more terminals.

Although the future of the newly built gas infrastructure beyond the next 20 years is unknown, the potential risk of stranded assets is not believed to be a major factor (Al-Kuwari, 2023). Investors in China are not currently concerned with infrastructure beyond the next 20 years, and LNG terminal owners and operators expect quick returns on investment. An expert in the development of China's gas market noted that stranded assets, underutilisation, and decarbonising natural gas infrastructure were not issues that would slow down infrastructure development given prolific profits and a recent history of positive market developments (Al-Kuwari, 2023). Even though many LNG terminal investment decisions were made before China's 2030 and 2060 dual carbon announcements, there is a belief that terminals will be built anyway. China's natural gas market has been a supply- and infrastructure-constrained market in the past, and this has not changed fundamentally. Despite room for growth for domestic production, international cross-border pipelines and imported LNG are needed. Nonetheless, China's LNG terminals and planned pipeline projects may meet demand soon, and spare capacity in LNG terminals could facilitate additional imports. In the early days of LNG imports, companies used various means to import and deliver LNG beyond receiving terminals, and produce and transport it by road. While LNG helps address infrastructure bottlenecks, it is not a quick solution given the time required to build the right infrastructure. Furthermore, China has challenged infrastructure investors in the past – as witnessed by the history of large companies operating oil and gas facilities, including Shell and BP. Interviewees agreed that LNG is favourable compared with upcoming trans-border pipeline projects given its proximity to coastal markets, ability to meet seasonal heating demand, and accessibility to LNG regasification infrastructure (Al-Kuwari, 2023). However, Chinese buyers will need to commit to long-term supply even when there is an oversupply in global LNG markets for LNG to become more affordable and stable.

Key drivers and tensions

Based on the interview analysis (Al-Kuwari, 2023), key variables and policy drivers affecting natural gas and LNG's future include energy security, the role of coal, climate change politics and commitments, central and regional government dynamics, and market liberalisation.

Energy security

In March 2021, the National People's Congress (NPC) approved the "Outline for the 14th Five-Year Plan (FYP) and Long-Term Targets to 2035", which devoted a section to energy security, stating that the core demand for oil and gas should rely on self-sufficiency (Downs, 2021; Meidan et al., 2021). China's shift to becoming an oil importer introduced a debate on whether to manage energy security by relying on the international energy market or by focusing on self-reliance (Downs, 2004). The 14th FYP resolved the debate by emphasising supply diversification via market liberalisation, facilitating diversity of suppliers, infrastructure development, and implementation of price reforms to improve system robustness and flexibility (Meidan et al., 2021). China successfully developed strategies to minimise its energy security risks through overseas investments, increased strategic petroleum reserves, diversification in the energy value chain and suppliers, and efforts to reduce waste (Andrews-Speed, 2015; Downs, 2004; IEA, 2023; Meidan, 2020; Moe & Midford, 2014; Vivoda, 2019). Overall, securing the supply of oil and natural gas to China remains "job number one" for the NOCs (Downs, 2021). In the 2019 National Gas Development Report (NGDR), energy security was highlighted as a main concern in the context of the gas industry's development. The report identified the main challenges for gas development in China – the increase in natural gas imports with the associated impact on energy security, the need to balance environmental protection with oil and gas production, and the lack of sufficient gas-storage facilities (O'Sullivan, 2019). Policymakers can use the need to reinforce energy security objectives as an opportunity to reshape the energy structure to make it less dependent on fossil fuel imports. At the same time, energy security also represents the biggest barrier to achieving China's climate targets – one of the main competing challenges (Al-Kuwari, 2023).

Domestic resources

The focus on self-sufficiency, together with the abundance of recoverable reserves of coal, has solidified the dominant role that coal plays in the energy mix in the context of energy security (Le et al., 2019; Zhang & Andrews-Speed, 2020). The availability of inexpensive domestic coal, historically favourable environmental regulations, large-scale coal-fired power generation capacity, and low retail electricity costs all underpin the consistent increase in coal consumption that has, in turn, supported rapid economic growth (Chandra, 2017; Zhongying & Sandholt, 2019; Zhou & Qin, 2020). This explains the Government's continued focus on the development of new coal mines, unconventional oil and gas resources, coal liquefaction, and gasification as measures to curb import dependence (Andrews-Speed et al., 2020; EIA, 2023). At the same time, large-scale coal consumption has contributed to severe air, water, and soil pollution, increasingly counteracting coal expansion in policymaking circles, reflected by the increased focus on coal-to-gas switching and aggressive development of renewables. Overall, though, China's coal-rich resource base is a barrier to the country's transition plans.

Another key barrier to the extensive development of renewables in China is an inability to ramp up in sufficient scale to meet the country's prodigious energy demand growth. There is an undeniable continued role for fossil fuels, mainly linked to the lack of large-scale storage facilities, which interviewees noted is a major barrier compared with natural gas use. One expert highlighted the winter of 2020 as demonstrating the limited role renewables can play in the current energy mix (Al-Kuwari, 2023). When China's economy picked up, a cap on coal consumption combined with supply shortfalls led to demand-side responses to mitigate power outages. The current energy-intensive economic model requires all forms of energy to fuel the economy. The current pace of renewables development will not fill the gap given the size and growth of demand. China's energy system requires a more sophisticated solution than can currently be provided by a wholehearted shift to renewables.

China's proven gas reserves are mainly located in the north-west – for instance, the Tarim Basin in Xinjiang, the Ordos Basin in north Shaanxi, southern Inner Mongolia provinces, and the Sichuan Basin in the southwest (Gao & Zheng, 2022). China also has substantial shale gas potential, with the Government using shale gas development to diversify energy supply (Le et al., 2019; Yunna & Yisheng, 2014). Improved subsidies, continued investment in research, and standards harmonisation could overcome some of the barriers affecting the development of shale gas in China (Ren et al., 2015; Yunna & Yisheng, 2014). The NGDR has focused on developing the Sichuan, Ordos, and Tarim basins, where most of China's conventional and unconventional gas resources are located. Overall gas exploration and production continue to grow, with shale gas delivering increased output of over 35% year on year from 2019 and 2020, indicating that government efforts to increase unconventional gas production are successful (O'Sullivan, 2019). If economy-wide switching from coal to natural gas is to be achieved, gas transportation to eastern and central regions will be necessary given natural gas reservoirs are most abundant in Western China (Xu & Lin, 2019). The Chinese Government, therefore, plans to ease restrictions on foreign and private investment in conventional oil and gas upstream projects to attract more investment and technical expertise (EIA, 2020).

Al-Kuwari's (2023) interviewees acknowledged some of the Government's successes in increasing shale gas production but noted that the increasing reliance on unconventional natural gas renders domestic natural gas extraction expensive and technically difficult, with coalbed CH_4 specifically creating pollution problems. Shale gas reserves are found in regions with high populations and limited water resources, making them difficult to develop rapidly. Chinese companies may claim shale gas development as a success story, but growth has been limited, with some industry experts concluding that, realistically, it cannot grow to support half the domestic gas demand. While local gas production plays a crucial role, a great increase cannot be expected.

Role of imports

In 2021, China imported 70% of its oil and 42% of its natural gas (Downs, 2021; IEA, n.d.). In April 2021, CNPC projected that by 2025 dependence on imported oil would range between 70% and 75% and natural gas between 34% and 66% (Downs, 2021). Increased dependence on imports in the medium term has led policymakers to adopt several strategies. For instance, the Ministry of Foreign Affairs developed relationships with oil-producing states to enhance energy security, and China's diplomats support overseas investments made by Chinese oil companies (Downs, 2004). Since 2010, natural gas pipeline imports from Central Asia, Myanmar, and Russia further contributed to long-term diversification efforts, with pipeline imports from Russia playing an increasingly important role (Vivoda, 2019). In May 2014, China's President Xi Jinping and Russia's President Vladimir Putin signed a US$400 billion agreement for Russia's Gazprom to supply 38 BCM of gas per year by 2018, followed by another 30 BCM per year over 30 years (Colombo et al., 2016). China's policymakers assessed pipeline projects from different supply sources by ensuring physical connections with Turkmenistan and Myanmar and additional LNG infrastructure, before moving forward with the Russian pipeline gas agreement (Colombo et al., 2016).

The Power of Siberia pipeline will connect Russia's Eastern Siberia reserves with China's North-eastern provinces, with the full capacity of 38 BCM per year expected by the mid-2020s (Sandalow et al., 2018). China views the gas deals with Russia as a means to diversify its import sources in the context of long-term energy security, and to manage emissions. Following the Russian invasion of Ukraine in 2022, China has become the largest single country market for Russian gas exports, with the war having a substantial impact on the global gas trade and significant geopolitical implications for the United States and the EU (EIA, 2024). However, strategic and technical considerations affect the role and scale of pipeline-imported natural gas; the potential for increased pipeline gas imports from Central Asia is limited. Turkmenistan has under-delivered on its 55 BCM per year capacity, partly because of its own growing domestic needs. Longer term, Kazakhstan and Uzbekistan may be able to ramp up gas volumes but it is unclear by how much (Meidan, 2019; Sandalow et al., 2018). Pipeline imports are vital given the need to use existing infrastructure, but there is uncertainty about whether China can increase its reliance on pipeline gas beyond its current plans owing to issues faced in recent years from Central Asian imports, especially in winter (Al-Kuwari, 2023). Pipeline gas deliveries from Turkmenistan fell substantially from the second half of 2017 because of greater domestic demand, cold weather, and unplanned outages (Sandalow et al., 2018) but became the largest source of gas imports to China in 2022.

The Belt and Road Initiative (BRI) was adopted in 2013 and involves 65 countries, accounting for approximately two-thirds of global recoverable oil and natural gas reserves (Zhao et al., 2021). The BRI enables China to secure gas supply by promoting energy investment and pipeline construction to support the diversification of import supply in associated countries (Lin & Li, 2020; Tan et al., 2021; Zhao et al., 2021). By 2018, 28 BRI countries were cooperating with the Chinese

Government or Chinese companies in exploitation of oil and gas resources, mining, international trade, and processing (Zhao et al., 2021). Cooperation with BRI countries introduces a risk, however, because many face political and social instability, economic turbulence, and significant differences in resource reserves. Zhao et al. (2021) assessed the risk to cooperation between China and BRI countries for oil and gas resources, and deemed Saudi Arabia, Russia, the United Arab Emirates, Qatar, and Singapore as particularly low-risk potential partners.

Role of LNG

Chinese companies are encouraged to secure resources abroad to improve diversity of supply through participation in LNG projects, partly by securing upstream supplies for the domestic market (O'Sullivan, 2019). The CNOOC was the first NOC to purchase an equity stake in LNG projects in Australia and Indonesia while entering into long-term sales agreements (Tsafos, 2019). In the early years, Australia represented over 80% of China's LNG imports, and in 2015 an increase in Australian imports further lowered China's diversification (Vivoda, 2019). Nonetheless, China has taken a long-term strategic approach to LNG market development, including proactive measures to secure LNG supply, in many ways adopting Japan's multiple-sourcing approach (Bradshaw & Waterworth, 2020; Colombo et al., 2016). China has steadily expanded its ties with Qatar, Indonesia, and Malaysia as part of its energy diplomacy to diversify its share of imports (EIA, 2023; Vivoda, 2019). Portfolio companies such as BP, Chevron, Engie, Petronas, and Total play a major role in delivering LNG and act as an extra buffer to security, reducing the need to depend on dedicated sources (Tsafos, 2019).

LNG versus pipeline gas

The FYPs do not mandate output goals of pipeline supplies versus LNG, but SOEs will likely be cautious and aim for a balance between the two sources (Meidan et al., 2021). New pipeline supplies from Russia compete with LNG imports from the north and support China's natural gas supply diversification (EIA, 2020, 2023; IEEFA, 2024). While the planned increase of pipeline gas imports from Russia could have a direct adverse effect on total LNG demand growth, in the long term Russia's eastern gas fields have limited ability to adjust seasonally, highlighting LNG's role as a flexible fuel source (Miyamoto & Ishiguro, 2018; Sandalow et al., 2018). LNG flows are likely to outpace additional pipeline arrivals while operating and regulatory guidelines are streamlined (Meidan et al., 2021). In the context of the perceived reliability of gas imports, the view that Central Asian gas imports have performed inconsistently in recent winters strengthens support for LNG from a seasonal demand perspective.

US–China relations will partially shape China's energy policies and have global ramifications on the global energy transition (Meidan, 2020). In October 2019, possibly in response to the trade conflict with the United States, Minister and Chinese Premier Li Keqiang commented that China would take advantage of its domestic resources, including coal, to bolster energy security (Andrews-Speed et al., 2020). One Chinese market expert noted that the US trade conflict makes securing supplies and upstream stakes more challenging for Chinese importers. Guo and Hawkes (2019) analysed the impact of a 25% tariff imposed by China on US natural gas, noting that, in the long term, the tariff will have a negative effect on the Chinese market as global supply tightens.

Natural gas is integral to energy security in China. In the past, China did not want to import more than 50% of its gas. However, domestic gas production has not kept up with demand, and this has led to increased LNG and pipeline gas imports. In addition, several interviewees concluded that both pipeline and LNG imports will eventually come under pressure if gas demand starts to plateau or decrease, compared with domestic production (Al-Kuwari, 2023). Under-investment in supply by the industry presents a great risk for LNG's future in China. High prices linked to supply pressure were noted as the biggest potential reason for LNG failing in China (Al-Kuwari, 2023). Another of Al-Kuwari's (2023) participants highlighted the winter of 2017–2018 when the average utilisation rate of the northern LNG terminals exceeded 120%. This increased demand led to natural gas shortages, driving up costs. Such events damaged the perception of LNG as a secure source. The lack of storage facilities was noted as another fundamental concern affecting consideration of LNG as a secure source. While CNOOC, for instance, plans to build considerable storage facilities, this would still not be sufficient to meet the country's growing demand.

More broadly, however, importing LNG and gas goes against the general security of supply agenda. It exposes China to volatile global markets and contributes negatively to trade balances with foreign countries. When compared to significant domestic coal reserves, LNG in particular is at a major disadvantage. One industry executive noted that importing more gas would establish an increased dependency, which does not support the political plans for the future use of gas. Thus, policymakers may decide "enough is enough" and aggressively explore other options (Al-Kuwari, 2023).

Role of coal

China's energy-transition journey is particularly challenging given its dependence on coal. Heavy coal use has led to extreme air pollution in Chinese cities, particularly in the north and north-east given their close proximity to coal reserves and operational coal-fired plants (Gilbert & Sovacool, 2017; Huang, 2024; Isoaho et al., 2017; Khan & Yardley, 2007; Weng et al., 2023). The ruling Communist Party of China (CPC) understood the potential threat that increased air pollution could pose to society, especially following the winter of 2012–2013 when air quality became so poor, it was dubbed the "airpocalypse", after which tackling air pollution became a high governmental priority (Downs, 2021). In 2013, the "National Action Plan on Air Pollution Prevention and Control (2013–2017)" was adopted with quantitative targets and policies to reduce particulate emissions and other pollutants. The plan recognised coal as a key air pollution driver and signalled a strong policy shift to limit its use and move towards other fuels (Meidan, 2020; Miyamoto & Ishiguro, 2018; Sandalow et al., 2018). In 2015, smog was so severe that Chinese authorities declared a red alert, closing schools, halting construction work, and limiting traffic movement (Gao, 2015). Poor air quality forced the Government to enforce fuel-switching away from coal in industrial use, power generation, and residential and commercial heating. Urban areas in north-eastern China were particularly targeted, where coal use in the winter causes smog and excessive pollution (EIA, 2020). President Xi proclaimed that he would "make China's skies blue again" (Sandalow et al., 2018), and China has met most of the goals laid out in the National Air Quality Action Plan. Populated areas have experienced significant air-quality improvements between 2013 and 2017 (Greenstone & Schwarz, 2018), and initial efforts to reduce SO_2 and NO_X emissions were successful owing to the installation of desulfurisation and/or de-nitrification facilities in large power stations and industrial plants. The continued reduction of $PM_{2.5}$ will be more challenging, however, because particulate matter is emitted from all sectors, including transportation (Huang, 2024; Miyamoto & Ishiguro, 2018).

NOCs are key to addressing air pollution and have so far installed expensive refinery upgrades to produce cleaner fuels, and increased natural gas production and distribution infrastructure (Downs, 2021). In 2018, the State Council published the "Three-Year Action Plan for Winning the Battle for Blue Skies" to address air pollution under the 13th FYP, an expansion of the 2013 action plan that called for an increased share of natural gas in the energy mix. This included expansion of production, supply, storage, and replacement of coal-fired boilers with natural gas units, especially in the Beijing–Tianjin–Hebei region (Downs, 2021; Isoaho et al., 2017; Luo et al., 2021; Miyamoto & Ishiguro, 2018).

Coal is not "unanimously viewed as the climate villain in China" (Meidan, 2019, p. 2). The coal industry is a major employer and key provider of tax revenue. Coal use is underpinned by a strong political base given its domestic production, large-scale employment, and energy security (Colombo et al., 2016; Meidan et al., 2021; You, 2022). Its overall role in supporting the economy is also evident in the way China controls coal import levels to support domestic producers by leveraging import quotas or implementing restrictions at entry ports (EIA, 2020). Future demand for coal will be determined by the growth in electricity and industrial needs, and by the Government's policies on coal switching and air quality (EIA, 2020). Coal has a strong advantage linked to the country's abundant reserves, accounting for 93% of recoverable domestic resources. Identifying pathways to use these reserves in exchange for imported oil and natural gas is a strategy policymakers have been considering for years (Li et al., 2020). Technology upgrades, CCS, operational improvements to existing plants, and enhancing operational loads could promote more clean and efficient use of coal.

Investment in coal supply in China was reshaped by the 2016–2017 reforms, which resulted in the closure of smaller mines that are less productive and efficient. After a two-year pause, the National Energy Administration (NEA) and the NDRC restarted approving new mine applications, accounting for 28 Mtpa of additional capacity in 2017, 68 Mtpa in 2018, and 201 Mtpa in 2019 (EIA,

2020; IEA, 2020). In 2019, despite an overall reduction in coal-fired powerplant investment, final investment decisions increased in China (IEA, 2020). Although decreasing in the primary energy mix, coal still dominates (Al-Kuwari, 2023; EIA, 2023). The Government appreciates its low cost and ability to stabilise end-users' cost. The power network and its heavy reliance on coal requires massive investment to shift away from a coal-dependent system, which requires strong and consistent political support.

Government efforts to switch away from coal in a sustainable manner have manifested in several ways. The Strategic Action Plan for Energy Development (2014–2020) set out a reduction in coal consumption and facilitated switching to other fuels (Miyamoto & Ishiguro, 2018). From 2017, an estimated 4 million households switched from coal to natural gas as part of China's first clean winter heating plan that required 28 cities to replace small coal-fired boilers with natural gas-fired units (Sandalow et al., 2018). Policymakers plan to eventually phase out coal from the residential sector for heating purposes as the Government targets household conversion from coal boilers to natural gas or electric ones (EIA, 2020).

There is significant potential for industrial coal-to-gas switching, for example, where it is used as a heat source in cement production and as a feedstock in chemical production. Accordingly, natural gas demand in the industrial and commercial sectors increased from 177 BCM in 2013 to 280 BCM in 2018 (Meidan, 2020). Opportunities also exist to transition from coal to natural gas for the production of ammonia and methanol, which would reduce energy intensity and process emissions (Romsom & McPhail, 2020). The Government plans to target industrial parks and energy-intensive industries to reduce carbon intensity and control energy intensity by supporting industrial coal-to-gas switching, especially in coastal areas. Stern (2017) concluded that air-quality problems and large-scale replacement of coal by gas-fired power generation may ultimately lead to natural gas carving out a different role in the medium to long term in China's energy system.

Imported pipeline natural gas tends to bring the most air-quality and carbon benefits, while imported LNG is superior to synthetic natural gas (SNG) – it does not have adverse impacts on water sources and improves air quality (Qin et al., 2018). However, emissions benefits will only occur if LNG is used to replace coal in power generation, as opposed to it displacing low-carbon sources or meeting incremental demand (Gilbert & Sovacool, 2017, 2018). In the short term, natural gas is clearly the best option for replacing coal in terms of scale and speed because gas-fired power plants can be constructed and commissioned quickly, whereas renewables and nuclear have longer lead times (Al-Kuwari, 2023). To illustrate the scope for gas demand growth, one participant calculated that replacing non-power-related coal demand in China would equate to the size of the current LNG market. LNG and gas provide a tangible opportunity for substantial and rapid displacement of coal given the Government's understanding of the impact of respiratory problems and that the current situation is untenable. In addition, continuing to burn coal contradicts the strategic goal of clean and green development. Utilities have an interest in cleaning up the electricity system, bringing in more nuclear and renewables and using less coal. LNG is cleaner and easier to store than coal, assuming storage facilities are built, and natural gas is more efficient and flexible than coal, giving it a clear technological advantage. In the future, non-power sectors could play a deciding role in the success of coal-to-gas switching, where air pollution issues are more substantial owing to limited emission treatment. Conversely, while coal use reduction is a top priority given air-quality concerns, since 2016 the country has occasionally loosened restrictions on domestic coal production to meet demand, demonstrating the Government's ability to switch back as needed should other priorities arise. One industry executive concluded that China is likely to develop coal until at least 2040, focusing on new, more efficient plants and increasing the application of CCS technology to replace existing, high-emissions power plants (Al-Kuwari, 2023). CCS also offers the Government the potential to develop clean coal, a potential medium-term competitor to natural gas.

China may not be affected by many constraints in the coming decade, and should continue to adopt a flexible, non-dogmatic approach to coal use. In terms of a clear pathway for the role of gas, China has been targeting gas growth for years, but the energy system is still unable to cope with demand peaks, increasing concerns regarding switching permanently and at scale. Coal-to-gas shifts are delayed when LNG is expensive, and a lack of gas when needed poses significant challenges, likely extending coal use and shortening the available time for gas as a transition fuel. Interviewees also concluded that, at some level, gas demand policy is uncertain. Despite initiatives related to air-quality improvements and gas infrastructure developments, there are no compelling signals of aggressively promoting gas usage.

Climate change politics and commitments

China's policymakers must reconcile competing priorities – environmental protection and climate change on the one hand, and economic growth, employment, and poverty alleviation on the other (Andrews-Speed et al., 2020). Environmental degradation and its impact on the population, mainly owing to high levels of air pollution and water contamination, mean that environmental protection is an energy policy priority from social necessity and industrial opportunity perspectives (Meidan, 2020). In the three FYPs between 2000 and 2015, coal and oil were promoted as the main fuels to support China's economic development (Zhongying & Sandholt, 2019). China's leaders were concerned about the rapid increase in the country's energy demand in the first decade of the 2000s and deemed the continuing reliance on fossil fuels particularly unsustainable (Downs, 2021). Thus, in 2017, at the 19th Communist Party Congress, President Xi committed to the development of "ecological civilization" as China's main priority and a prerequisite for economic development (Zhongying & Sandholt, 2019).

Disincentivising coal-based power, improving renewable energy integration, developing electric vehicle infrastructure, reforming the power sector, and setting up an ETS are policy measures taken by the Government (EIA, 2020; Zhongying & Sandholt, 2019). Efforts to reduce energy intensity have been successful owing to the Government's use of administrative instruments, including cascading targets for government entities and SOEs. China reduced its carbon intensity by 46% by the end of 2017, compared with 2005, fulfilling its 40–45% carbon-intensity reduction target three years early (Andrews-Speed, 2015; Wen et al., 2020).

2030 and 2060 dual carbon goals

China firmly committed to the Paris Agreement, pledging its CO_2 emissions would peak and cutting carbon emissions per unit of GDP by 60–65% from 2005 levels by 2030 (Koyama, 2020; Sandalow et al., 2018). On 22 September 2020, President Xi announced that China would raise its NDCs, aiming to reach peak CO_2 emissions before 2030 (EIA, 2023; Liu et al., 2023). Additionally, by 2060 CO_2 emissions per unit of GDP would decline by 65% from 2005 levels, and the share of non-fossil fuels in the energy mix would rise to 25% by 2035 (Andrews-Speed et al., 2020; Kaaresvirta et al., 2021; Koyama, 2020; Liu et al., 2023; Meidan et al., 2021). The President also announced that China would seek to achieve carbon neutrality by 2060 (Kaaresvirta et al., 2021; Liu et al., 2023). This strengthening of China's climate change commitments served to enhance its role as a responsible global power, increasing its international community presence (Koyama, 2020).

With lower levels of economic growth, increased availability of renewables, the steady decline of heavy industry, and expanding service sector, it is likely that China's CO_2 emissions will peak before 2030 (Andrews-Speed et al., 2020; EIA, 2023; Kaaresvirta et al., 2021; Liu et al., 2023; Wen et al., 2020). Zhang and Andrews-Speed (2020) concluded that the planned reduction in carbon emissions will be achieved mainly through reduced emissions in those industrial sub-sectors where China's pilot carbon trading scheme operates. It will be much more challenging, however, to reach carbon neutrality. To do so, nuclear, renewables, hydrogen, energy storage, fuel cell, CCS, and industrial electrification all must be developed at high rates or completed alongside overhauling the energy industry and industrial and transport sectors (Andrews-Speed et al., 2020). Kaaresvirta et al. (2021) concluded that to reach carbon neutrality by 2050, China would need renewables and nuclear to meet 85% of its energy consumption and more than 90% of its power generation. Solar energy would need to increase tenfold, and wind and nuclear sevenfold from current levels. To fully capture the climate benefits of switching away from coal, China must also focus on detecting and preventing CH_4 leaks while developing its natural gas infrastructure (CarbonBrief, 2023; Sandalow et al., 2018). Understanding the importance of this, the country joined a coalition of importing countries – including South Korea and Japan – and the EU to coordinate efforts on energy-related emissions as part of the EU Methane Strategy (Stern, 2022).

China must also seek to decrease the intensity of GHG emissions from its overall gas supply, including LNG imports. Gan et al. (2020) estimated the well-to-city-gate GHG emissions of China's gas supplies. Supplies from Russia's Urengoy and Nadym fields, Turkmenistan's Galkynysh and Bagtyýarlyk fields, and domestic shale gas from the Fuling field, all have greater than the 75th percentile well-to-city-gate GHG emissions intensities at the supply curve (Gan et al., 2020). The GHG emissions intensity of supplies is also set to increase as more GHG emissions-intensive gas supplies are sourced from Russia, Central Asia, and domestic shale gas fields (from 21.7 CO_2 equiv MJ-1 to 23.3 CO_2 equiv MJ-1 in 2030) (Gan et al., 2020).

Central and regional government dynamics

Government system and structure

The legal and regulatory systems are seen as government instruments that the CPC uses to retain power, maintain social order, and promote economic development. There are three levels of government – central, provincial, and municipal (Zhang & Andrews-Speed, 2020). Political power and authority lie primarily with the CPC, and then with the Government and NPC. President Xi has been the CPC's Party Chairman since 2012 (Zhang & Andrews-Speed, 2020). China has a political economy and a legal system that reflects a coordinated market economy, with energy policy and regulations formulated in a state-centric paradigm where the state owns energy companies, controls pricing, and applies policy instruments to modify behaviour (Andrews-Speed, 2015, 2016). The energy sector has historically been fully government-controlled since the advent of the Marxist–Leninist system in 1949 (Zhang & Andrews-Speed, 2020; Zhou et al., 2021). Deng Xiaoping's strategy of progressive liberalisation transformed China's economy after 1978, moving away from the highly centralised process coordinated by the State Planning Commission (Zhang & Andrews-Speed, 2020). The state, however, remained in control of strategic sectors, defined as "pillar" industries, including banking, energy, telecommunications, mining, metallurgy, chemicals, and rail.

In 2008, the NDRC – in charge of China's macroeconomic and social development – was granted additional power covering energy, climate, and carbon reduction policies, and has energy planning and pricing oversight (Isoaho et al., 2017; Meidan et al., 2021; Zhang & Andrews-Speed, 2020). The NDRC is responsible for overall strategy and planning, pricing, fiscal and macroeconomic policy, reform, and major project approval (Rioux et al., 2019). The NEA was established under the NDRC in 2008, and acts as the de facto energy ministry, focused on industrial policy and supervision (Isoaho et al., 2017; Meidan et al., 2021; Rioux et al., 2019). The National Energy Commission (NEC) was created in 2012 to streamline energy policy decision-making and to establish the national energy development strategy (Isoaho et al., 2017). The NDRC and CNEA produce FYPs, controlling energy prices and overseeing energy policy implementation (Zhang & Andrews-Speed, 2020).

The Ministry of Natural resources (MNR) defines policies for mineral rights allocation, exploration of resources, and foreign cooperation in the upstream sector (Rioux et al., 2019). In March 2018, the Ministry of Ecology and Environment (MEE) was established with a broad remit to reflect the importance of environmental issues and improve coordination between the SOEs (Sandalow et al., 2018).

The MEE replaced the Ministry of Environmental Protection (MEP) to improve the effectiveness of regulation by clarifying bureaucratic responsibilities and reducing duplication (Meidan et al., 2021). The MEE takes the lead on climate policy, highlighting its efforts to meet the 2030 peak carbon goal, and implementing a range of initiatives, including the ETS (Meidan et al., 2021). The Department of Climate Change Mitigation at the provincial level, meanwhile, allocates emissions allowances to entities within each jurisdiction (Zhang & Andrews-Speed, 2020).

SOEs

The CPC and SOEs have a strong relationship that shapes China's energy and power sectors. Parent companies are among the 97 SOEs under the administration of the State-Owned Assets Supervision and Administration Commission (SASAC), which reports directly to the State Council (Downs, 2021). The energy sector, including oil, gas, and other natural resources, remains one of a few strategic sectors in which government control continues, mainly via corporate strategy and senior appointments (Andrews-Speed, 2015). Energy is identified as strategically important, with oil and gas being pillar industries and wind and solar being new energy technology industries requiring direct state intervention (Andrews-Speed, 2015). In recent years, government involvement and control have been exercised through state-owned energy companies, initially through the Ministry for Petroleum Industry for coal and electricity (Andrews-Speed, 2015). SASAC reviews the performance of SOE executives, considering criteria such as net profits and economic value added (Downs, 2021).

China established NOCs to manage natural gas production and transportation infrastructure, established a licensing system, and allowed private funding of gas exploration (Rioux et al., 2019). In 2020, the three government-owned NOCs – CNPC, China Petroleum & Chemical Corporation (Sinopec), and CNOOC – and their flagship subsidiaries represented more than 90% of production of China's oil and natural gas – 3.9 million barrels per day and 192.5 BCM, respectively – as well as 60% of China's refining capacity (Downs, 2021). The CPC appoints these NOCs' top executives. They also belong to a group of 50 core SOEs contributing to strategic alignment with government policy priorities.

In recent decades, the NOCs increased their efforts at energy efficiency and air pollution reduction to align with the country's priorities of slowing energy demand growth and improving air quality (Downs, 2021). CNPC and Sinopec are among the highest GHG emitters among publicly traded companies, partly owing to large upstream and downstream operations, while CNOOC has significantly lower emissions per barrel of oil equivalent produced and gross emissions, partly linked to the lack of operating refineries in its portfolio (Downs, 2021). In May 2021, CNPC, Sinopec, CNOOC, and four other Chinese oil and natural gas companies (Beijing Gas Group Co., Ltd., China Resources Gas Group Ltd., ENN Energy Holdings Ltd., and China Oil and Gas Pipeline Network Corporation) established the China Oil and Gas Methane Alliance with the aim of reducing the average CH_4

intensity of natural gas production to below 0.25% by 2025 from an unidentified level (Downs, 2021; IEA, 2024).

NOCs are subject to market and domestic regulatory pressures (e.g. SASAC set profit targets), impacting their climate change mitigation efforts (Downs, 2021). In August 2020, PetroChina announced a target of net-zero emissions by 2050, and in March 2021, PetroChina and Sinopec declared their aim to reach peak carbon emissions by 2025, with Sinopec intending to be carbon neutral by 2050 (Downs, 2021). CNOC plans a 16% reduction in GHG emissions between 2020 and 2025, peak carbon emissions before 2030, and carbon neutrality by 2060 (Downs, 2021). The NOCs are thus on a pathway to engage with climate change.

Power structures and their impact on energy policymaking

The Government has demonstrated an ambition for a structural shift with recent governance changes, including criticism of CNEA relating to environmental protections and the elevation of the NDRC's role in energy policymaking (Meidan et al., 2021). In February 2021, The Central Environmental Inspection Team (CEIT) criticised the CNEA's expansion of coal-fired power and lack of promotion in developing low-carbon energy (Meidan et al., 2021). This public display was noteworthy because of its explicitly critical nature of a central government agency and the fact that it was made by an implementation group directly under the CPC (Meidan et al., 2021). Interviewees agreed that the Government, in its different forms, is the main actor in driving and influencing energy policy (Al-Kuwari, 2023). More specifically, the NDRC and State Council are the main bodies shaping the energy transition, aided by coordination between the NOCs and SOEs and the actions of the CNEA as regulator.

The CNEA's Advisory Experts Committee is the most influential expert body under the NEA, and its purpose is to support national energy development plans and policies. Zhou and Qin (2020) highlighted that nine of the committee's ten members worked directly with the NOCs, either currently or previously. Top SOE and NOC executives generally hold vice-ministerial rank and are "appointed, removed, and promoted by the Central Organisation Department (COD)" (Downs, 2021, p. 11), the CPC's human resources arm. Career prospects of those executives are impacted by how state interests are advanced (Downs, 2021). The promotion of SOEs has limited the role of private companies, and contributed to corruption, abuse of market power, and low levels of innovation (Zhang & Andrews-Speed, 2020). SOEs have strong ties to the political elite, which could limit the effectiveness of new measures, such as the carbon ETS (Andrews-Speed, 2016). SOEs have also shown the ability to resist initiatives, for example, by making the transmission grid system less favourable to renewables (Isoaho et al., 2017). SOEs in strategic sectors, such as power generation, have majority state ownership and receive special treatment to support their growth, including access to finance (Isoaho et al., 2017; Zhang & Andrews-Speed, 2020). In terms of natural gas, the NDRC designs policy, the NEA implements it in the form of FYPs for national energy development, and the SOEs construct pipelines (Gao & Zheng, 2022). Competition among the SOEs and NOCs has increased in terms of expanding their ability to lower the cost of LNG and increase downstream gas sales to ensure growth of their customer base. The CNPC has been identified as the NOC providing most of that support, dominating the landscape based on its long-term role of supplying 75–80% of domestic gas, causing other NOCs to be outsiders.

Central versus regional government

Local authorities have significant autonomy over economic development and land use rights (Isoaho et al., 2017). In fact, local governments can veto measures that could undermine central government policies, at times resulting in uncoordinated policy implementation (Zhang & Andrews-Speed, 2020). SOEs and local governments can also collude to undermine certain central government policies, as happened with the proposed construction of additional coal-fired plants between 2013 and 2016 (Zhang & Andrews-Speed, 2020). In recent years, central government has given state local authorities more power to allow provincial governments to approve new coal-fired and renewable power projects, leading to some economically motivated decisions on account of the local leaders' short-term focus (Isoaho et al., 2017). Local authorities will be allowed to apply stricter rules to allow for differentiation between regions (Zhang & Andrews-Speed, 2020). Provincial governments will also likely be given monitoring, reporting and verification, and accreditation responsibilities, risking inconsistency in applying standards across the country (Zhang & Andrews-Speed, 2020).

Market liberalisation

A gradual price liberalisation began with non-residential customers in 2015 (Sandalow et al., 2018). In 2017, the Government announced that third parties could access pipelines and LNG terminals. Foreign companies have been able to invest in natural gas city distribution pipelines since 2019, a reform intended to facilitate greater investment and faster infrastructure development (EIA, 2020). Energy sector privatisation and market liberalisation accelerated between 1997 and 2013, particularly under President Xi. Zhang and Andrews-Speed (2020) concluded that efforts to implement market mechanisms in the energy sector will face challenges from the institutional environment. The 2015 electricity market reforms proposed a set of measures, including the promotion of competition in power generation by allowing generating companies to negotiate with large customers, a system for regulating transmission and distribution tariffs, and the introduction of competition tariffs in electricity retail (Zhang & Andrews-Speed, 2020). Today, approximately 90% of generating capacity and nearly 100% of transmission and distribution capacity are controlled by central or local government (Zhang & Andrews-Speed, 2020).

Historically, the case for the privatisation and liberalisation of energy markets was economic and political. Through liberalisation, the Government's role has evolved from vertical command and control of policy formulation to the regulation of horizontal relationships (Zhang & Andrews-Speed, 2020). However, the lack of a strong independent regulator and the ability of local governments to interfere and grid companies to limit the emergence of new companies weaken the effectiveness of market reforms.

China lacks a specific plan to integrate its natural gas pipelines into the national distribution network. The objectives of these reforms may not be limited to the removal of subsidies, and multiple factors from assuring energy access to ensuring environmental sustainability may figure in decision-making (Lin & Li, 2020). The approach of a pipeline network comprising China Oil & Gas Pipeline Network Corporation (PipeChina) and local network companies does not necessarily lead to a reformed gas market (Li & Li, 2021). Market access and pricing mechanisms remain under government control, despite the progress made in market liberalisation (Rioux et al., 2019). PipeChina was created in 2020 to establish a national oil and gas pipeline network company and promote the construction of a nationwide network (O'Sullivan, 2019). The company changed gas trading to a multi-source, multi-customer, competitive mode pattern (Li & Li, 2021). PipeChina's initial operations faced challenges following a fire at Beihai LNG terminal and winter shortages. However, it enabled third-party access and launched a platform for shippers and LNG companies to register to apply for pipeline transmission (Meidan et al., 2021). PipeChina's operating philosophy is to liberalise imports and utilisation, and become a centralised intermediary infrastructure operator – an anti-monopoly philosophy with respect to NOCs. Liberalisation encourages LNG imports, mainly owing to third-party access to LNG terminals, which will unlock demand for downstream and some wholesale gas users.

The Government's control of natural gas prices leads to distortions and cross-subsidies across China, and prices do not accurately reflect natural gas demand (Lin & Li, 2020; Rioux et al., 2019). In fact, demand has grown despite the high gas import price linked to lower, distorted domestic prices. This has had a negative financial impact on the Government, and led to the 2013 natural gas reforms (Lin & Li, 2020). However, the natural gas industry continues to face difficulties. The "big three" have a high supply oligopoly, and links between upstream activities and the pipeline and storage and distribution networks continue to distort the market price (Luo et al., 2021). The Government's energy policies are increasingly moving towards market-based pricing schemes, energy efficiency and anti-pollution measures, competition among energy firms, greater access for smaller independent firms, and higher investment in shale gas and renewable energy projects (EIA, 2020; Zhang & Andrews-Speed, 2020).

Pricing

Pipeline gas price reform does not necessarily mean the removal of subsidies as a result of other competing priorities (Lin & Li, 2020). Wang et al. (2020) concluded that LNG import prices are affected by crude oil prices, and by total gas import volume and Henry Hub gas prices. Upstream exploration is open to foreign and private companies, and third-party access to LNG terminals on the coast introduces various resources (Li & Li, 2021). Liberalisation of the LNG market is driven by the Government's desire for a more flexible, price-competitive natural gas market (Bradshaw & Waterworth, 2020).

Recent market reform developments included the establishment of the Shanghai LNG Price Index by the Shanghai Petroleum and Gas Exchange (SHPGX) in 2015 in an effort to improve natural gas security of supply, improve China's global market bargaining power, and establish a regional benchmark price (Zeng et al., 2020). The SHPGX's strength is its ability to trade in both LNG and pipelines, which creates the potential for a future liquid price index (Bradshaw & Waterworth, 2020). The SHPGX LNG Price Index will have greater influence on market trading decisions once liquidity in the hub increases and further LNG reforms are implemented. However, Zeng et al. (2020) concluded that while the LNG Price Index is starting to exhibit long-term equilibrium and short-term adjustment mechanisms, market information transparency and price discovery efficiency remain inadequate.

Carbon trading

In 2021, China launched its national ETS, implementing President Xi's 2015 Paris Agreement announcement (Liu et al., 2023; Meidan et al., 2021). The full scheme began on 16 July 2021, initially covering 2,200 firms in the thermal power sector. Participants include 31 combined power PetroChina and Sinopec Corp subsidiaries. Companies that participate in the ETS programme receive carbon emissions permits at no cost, and those with lower emissions can sell their permits to companies with higher emissions. There are currently no auctions for permits, but the price of carbon permits opened at US$7.42/tonne and reached a high of US$8.18/tonne on the first day of operations (Downs, 2021). The ETS programme aims to reduce emissions per unit of economic output rather than focusing on total emissions. Therefore, while energy production may increase, companies can still increase their absolute emissions as long as they decrease emissions per unit of output (Downs, 2021). The carbon market is thus unlikely to curb emissions in the near term.

China's ETS will cover power generation and industry, including steel, iron, cement, and paper production, representing 80% of the country's annual CO_2 emissions, and estimated to be the largest carbon market in the world (three times greater than the European carbon market) (ICAP, 2022; Liu et al., 2023; Meidan et al., 2021; Sandalow et al., 2022). The ETS will initially cover 2,225 coal- and gas-fired plants. China's ETS is not based on the typical cap-and-trade schemes found globally, but on an intensity target (rather than an absolute cap) (ICAP, 2022; Liu et al., 2023; Meidan et al., 2021; Sandalow et al., 2022). Trading volumes and prices since the 2011 pilot ETS have been low – between €1/tonne and €15/tonne (Meidan et al., 2021). Key questions remain regarding transfers between the pilot and full ETS schemes, the eligibility of China's certified emissions reduction (CCER), and the management of carbon market regulation within the legislative hierarchy (i.e. State Council or MEE) (Meidan et al., 2021). Around 2,225 coal-fired power plants are covered by China's ETS market, and in the next two to three years, emissions-intensive industries like steel, cement, chemicals, oil, and gas will be added.

Carbon trading and carbon taxation may continue to face challenges. Large SOEs have significant market power, low budgetary constraints, and close relationships with local and central government. Governing agencies have limited capacity and authority to monitor and manage participants and offenders. Energy prices are broadly under government control, and the Government will be reluctant to allow a price increase that would be passed on to consumers (Andrews-Speed, 2015). The carbon ETS pilot projects had some drawbacks, including an overallocation of allowances that led to low carbon prices, low market liquidity, and poor monitoring, reporting, and verification (Zhang & Andrews-Speed, 2020).

Impact on LNG

In the short term, uncertainty as a result of slower economic growth, the impacts of PipeChina, and price reforms following liberalisation may make potential importers more cautious about gas demand certainty (Meidan, 2019). The Government has endorsed the concept of market-based pricing, which means that the market is ultimately moving in the right direction. There are already examples of LNG imports being more competitive than pipeline gas, even for some inland buyers, where swaps are being explored to take advantage of lower-priced LNG. This creates greater flexibility and more options to inland gas buyers. Clarity in the regulations and access rules, and more physical capacity will support the LNG market.

The market is playing a role, with CNPC and others being challenged, especially by PipeChina. LNG terminal slots are offered to outsiders and private companies. Commenting on the Shanghai LNG market exchange for LNG, a participant commented on its early stages, "It's not the UK yet. It's not because we're talking about this, that it's a free market like northern Europe". Another noted: "It is unclear how to fully liberalise. We are still in the trial-and-error phase". When the Shanghai LNG exchange first started trading, it was very superficial, with CNOOC posting only completed transactions (Al-Kuwari, 2023).

A core constraint is the lack of a market-based pricing system because pricing is currently linked to either oil or other hubs that do not reflect Asian supply and demand. This affects the ability to handle demand and supply shocks and encourages price volatility. In addition, some prerequisites are required to develop a real gas pricing hub, where price discovery takes place and liquidity is developed, which will require policies to liberalise the gas market and adapt the financial sector. In the past, users complained that CNPC held on to gas so that the winter price rises allowed it to benefit. As one stakeholder summarised:

> So you have all these debates – which I do not think [are] over despite the efforts, the good efforts – to try to liberalise, to optimise, to become more efficient, to vary the supply sources, and to increase storage, in order to avoid peak prices.
>
> (Al-Kuwari, 2023)

Non-NOCs have played an increasingly important role in the Chinese LNG landscape, signing long-term contracts and building receiving terminals. China is opening the LNG market to local organisations, including smaller SOEs and private companies. ENN, for example, is a private independent company that owns and operates its own terminal instead of going through an NOC, which would have been economically prohibitive. Downstream companies are building regasification terminals and city natural gas networks to cover "the last five miles" of the business, delivering gas to the end-user, while industrial organisations are looking to expand and include more added-value services and businesses to better leverage existing infrastructure. Opening up of the market also introduces risks. For LNG producers, the risk is the creditworthiness of new entrants. How contracts are structured and dealt with requires change from those with NOCs. Smaller players may also introduce health and safety and environmental risks and the chain and risk management is crucial to maintaining safety within the industry. In addition, a lack of sophistication by new gas buyers from the large NOCs may have led to bad deals where coal-to-gas switching could be hampered further, leading to smaller companies paying for a lack of flexibility.

Energy-transition dynamics

MLP analysis

Landscape analysis – social trends affecting the Chinese energy system

Chinese policymaking is centred on meeting people's needs, particularly given the improved access to information among its citizens. Following a period of rapid population growth, China's population has been ageing and shifting in structure over the past 30 years. People's awareness and expectations of their quality of life are changing and this affects China's energy transition. With more limited people in the workforce, it will be difficult to develop large-scale infrastructure projects. Substantial infrastructure is required in the next 20 years and must be developed rapidly if change is to occur. Chinese society has also become accustomed to certain luxuries. For example, SUVs are considered a status symbol, making the transition to low-emission alternatives unlikely in the short term (IEA, 2020). A participant from Al-Kuwari's (2023) study noted: "Now that people can afford to buy a car, the Government will have a difficult time forcing them to use electric cars".

Air quality is a significant concern in China. In 2011, the US embassy in Beijing issued tweets regarding $PM_{2.5}$ emissions, which attracted global attention and contributed to the political discourse (Gao, 2015; Miyamoto & Ishiguro, 2018). In 2012, <1% of China's 500 largest cities met the WHO's air-quality standards (Isoaho et al., 2017). The "airpocalypse" in the winter of 2013 gained widespread attention across Chinese society (Sandalow et al., 2018).

A study of public opinion presented at the World Economic Forum in 2020 showed that Chinese respondents have high levels of trust in climate science. In that same study, 42% of Chinese respondents considered climate protection to be more important than economic growth, comparable to UK respondents (44%) (World Economic Forum, 2020). In the spring of 2015, 76% of people surveyed in China stated that air pollution was a big problem, with 35% regarding it as a "very big problem" (Gao, 2015). China's leadership understands that societal pressures regarding emissions and air quality could destabilise the political landscape (Isoaho et al., 2017). In 2017, at the 19th Communist Party Congress, President Xi committed to "ecological civilization" as the main priority for China and a prerequisite for the country's economic development (Zhongying & Sandholt, 2019).

The shift from carbon-intensive resources to low-carbon systems requires major changes in employment scenarios, which could affect economic growth. SOEs will have to balance efforts between meeting the dual carbon goals, energy security responsibilities, and maintaining large workforces (Downs, 2021). Shifting employees from oil and gas activities to low-carbon jobs will require time and likely cause unemployment; CNPC and Sinopec especially have a "responsibility to keep people employed" (Downs, 2021, p. 35). In 2021, CNPC employed more than 1.2 million people, Sinopec Group over 0.5 million, and CNOOC around 80,000 (Downs, 2021). During the COVID-19 pandemic, the Government directed SOEs to hire more workers to support social stability and the economic recovery, in line with its commitment to job creation and societal betterment (Downs, 2021).

Innovation and the R&D ecosystem

China underwent a massive transformation over a 20-year period, moving from a technology importer to a major manufacturer and exporter of low-carbon technology, making it the second-largest investor in R&D (IEA, 2020; Wood Mackenzie, 2024). Global power sector investment was approximately US$760 billion in 2019, with China accounting for more than a quarter of overall investment (IEA, 2020). The public State Grid of China utility company (which accounts for around a third of electricity investment) announced investments totalling ¥450 billion in 2020 (around US$65 billion), with ultra-high voltage projects accounting for 40% of total investment – approximately 8 GW of coal-fired capacity was approved, similar to 2018 and 2019 despite the negative impact of COVID-19 (IEA, 2020). China accounted for three-quarters of PV module shipments in 2020 (Sandalow et al., 2022). The low-carbon component of energy-related R&D grew by 10% in 2019, with particular increases in energy efficiency and hydrogen.

However, the country's low-carbon investments have not kept pace with electricity consumption growth (IEA, 2020). China also mines rare earth minerals, and has been leading global production since the 1990s (Meidan, 2020). The 2008 financial crisis reduced global demand for solar PV products, leading the Government to respond by launching a ¥4 trillion stimulus package for emerging industries, including solar PV. Programmes to encourage the large-scale adoption of domestic PV were launched, including a FIT in 2011 (Andrews-Speed, 2015). However, the "Made in China 2025" strategy, an innovation plan focused on quality and new technologies, identified renewable energy and electric vehicles as strategic emerging technologies (Sandalow et al., 2022; Zhongying & Sandholt, 2019), confirming the energy sector's importance for decades to come.

China is one of the largest countries with hydrogen fuel cell technology targets for residential use, fuel cell vehicles, and refuelling stations, with the aim of developing ss MWe of hydrogen by 2021 (IEA, 2020, 2022; Sandalow et al., 2022). Fuel cells are seen as tools to address air-quality concerns and generate economic growth through their manufacture (IEA, 2022; Staffell et al., 2018). China is one of the largest potential markets for hydrogen imports from Australia (ACIL Allen, 2018), and its coastal industrial clusters could be part of the near-term scale-up opportunities to lower costs and create lower-carbon hydrogen hubs (IEA, 2019). In 2019, electrolysers were 83% cheaper to produce in China than in western countries (Bloomberg, 2020). Despite policymakers' interest in hydrogen, Al-Kuwari's (2023) interviewees expressed some scepticism in the role it could play in the near future. The lack of an economic model and the R&D funding that hydrogen receives compared with batteries or renewables was cited. Green hydrogen, although a key technology, may have particular difficulty competing with natural gas owing to cost and natural gas's significant infrastructure development plans. Fossil fuel–based hydrogen production could be attractive for China, given its access to suitable geological resources for storing captured carbon, subject to affordable fossil fuels (Bloomberg, 2020).

Considerable research has gone into the production of coal-based SNG in the past decade, particularly aimed at the northern coal region where CH_4 is extracted from coal for use in heating during winter. However, SNG accounted for only 2% of gas production in 2018 (Kaaresvirta et al., 2021). CCUS is key to meeting targets to address a coal-heavy energy structure, although serious plans have yet to materialise, and it is still in the R&D and technology development phases. Finally, advancements in LNG tanks and regasification unit cost improvements could support LNG in China.

Transition context

Energy-transition ambition

China's energy policy differs from that of other major countries because its energy transition has been primarily focused on air pollution. However, the country's profile as the world's top carbon emitter has garnered increasing scrutiny. Regardless of the drivers, the need to balance economic development with the pace of the energy transition will always influence plans.

In Al-Kuwari's (2023) study, participants noted that "the beauty of working with China is you have much more clarity than in other countries". Previously, despite the FYPs, there was a lack of clarity even though air pollution and coal switching were generally encouraged. However, targets will become mandates that make them "political and economic objectives from [the] top leadership". Another interviewee noted that no one is questioning the target, and the enforcement of environmental policies has become "more and more streamlined and more effective".

Climate targets were well received and "everyone is pushing" to meet them before 2060". Interviewees were generally positive that targets will be met. Given that the idea of achieving carbon neutrality by 2060 is politically well received, it is more likely to be implemented. Presidential mandates are very top-down. Peak and carbon-neutral targets achieved consensus across different government departments and sectors. While 2060 will not significantly affect NOCs significantly in the short term, the targets were well received. Despite clarity of vision, there is a lack of understanding on how these 2030 and 2060 targets will be achieved, which will have a major impact on the role of LNG in China. Experts and policymakers are currently trying to determine different paths. There is a joint committee-type effort to establish the targets because there is no single entity or organisation sufficiently powerful to have a good overview of the balance between decarbonisation and economic growth.

Adaptive capacity – energy capabilities and infrastructure

Multiple companies, both private and public, have the necessary expertise to improve energy capacity and infrastructure. NOCs have financial, technical, and management capability. For example, CNPC merged its oil and gas division with a new energy division in 2021, and plans to increase investment into renewables and hydrogen. The company leads in the deployment of CCUS with an operational capacity of 600,000 tonnes of CO_2 per year (Downs, 2021). Similarly, CNOOC has significant experience of offshore engineering and is well positioned to develop offshore wind having been a pioneer in this sector since 2007. However, China lacks the technology for complex reservoir exploration and production (Jia et al., 2021).

China could successfully exploit economic opportunities around decarbonisation, thereby continuing to drive down the cost of electric vehicles and other clean technologies. The country is largely self-sufficient in nuclear technology, having developed reactors and construction techniques for nuclear powerplants (Liu et al., 2023). Most reactors in China are Chinese-made, often with design and cooperation from foreign partners for certain technological elements (Kaaresvirta et al., 2021; World Nuclear Association, 2024). China plans to share this technology with countries such as Pakistan, Romania, and Argentina (Kaaresvirta et al., 2021; World Nuclear Association, 2024). The focus on self-reliance and distrust of international markets shapes the Chinese Government's paradigm regarding the low-carbon transition (Andrews-Speed, 2016). However, the Government must be mindful of time constraints. Participants in Al-Kuwari's study (2023) noted that "If . . . they don't promote the switch fast enough, then it is going to be even more difficult in the future because, economically, their best days are not in front of them".

China is on a transition path towards a new energy-generation mix. It has a rapidly evolving renewable energy deployment, FYPs with emissions-reduction targets per unit of GDP, and a pledge to increase the share of non-fossil fuels in primary energy consumption (Chapman & Itaoka, 2018). This transition offers the opportunity to leverage new technology to move away from fossil fuels, while simultaneously generating new avenues of economic growth. According to Grübler (2012), the countries representing the "energy periphery", led by China, may have the ability to transition away from fossil fuels faster than OECD countries that may experience lock-in through their entrenched technologies, lifestyles, and infrastructure.

Actors

The key actors broadly influencing the development of China's natural gas sector are the Government and related entities around its administration, the major integrated utility and gas companies, financial institutions, and large trading houses. Other influencers such as the media, end-users, and NGOs have a small but growing role.

Summary of the T-E-T-I analysis

This case study highlighted the factors that determine LNG's potential future pathways within China's energy system in the context of the current energy transition. Figure 5.2 summarises the T-E-T-I analysis for China. For the Government, LNG could provide another tool to switch away from coal to improve air-quality conditions. Tensions between the central and regional authorities owing to socio-economic conditions could lead to LNG potentially playing a different role

Resources

- LNG competes with domestic gas and pipeline imports; domestic gas production, including shale, to meet large part of demands
- Significant coal reserves; plans to ramp up nuclear in coming decades
- Renewables and nuclear demand cannot meet growing demand
- LNG provides seasonal option and coal-to-gas switching opportunities for power generation; close to demand centres on east coast

Market

- Market liberalisation opens access to new counterparts for gas supply and imports
- Financial affordability and the availability of sufficient global supplies impacts LNG attractiveness
- Some regions more prone to LNG development, depending on geography and economic profile
- Chinese ETS and carbon trading could support gas growth in near term
- PipeChina will support change in gas trading to multi-source, multi-customers, and be more competitive to optimise imports

Energy-transition Dynamics

- China is world's largest emitter of CO_2 and GHGs since 2006
- Ageing population and population structure rapidly changing in past three decades
- Air pollution initial impetus for decarbonisation
- China 2030/2060 dual carbon goals usher in new role in climate discussions
- 'Clean coal' R&D will continue to maximise resources
- Energy security and affordability linked to the support of economic development still take precedent over decarbonisation

Power

- Government restructuring of energy policymaking and sector leads to increased alignment in implementation
- SOEs aligned with government; also have ability to influence policy and plans owing to knowledge base
- Coal use flattening, and share looks to decrease; industry is a large employer and politically strong (not only a 'climate villain')
- BRI creates potential links with natural gas and LNG exporters
- US–China relations and impact of current and future trade issues

Figure 5.2 China case study following T-E-T-I framework analysis

based on geography and priorities. Affordability and the need for additional infrastructure nationally will ultimately be determinants in the role of natural gas, and consequently that of LNG.

References

ACIL Allen Consulting for Arena. (2018). *Opportunities for Australia from hydrogen exports.* https://arena.gov.au/assets/2018/08/opportunities-for-australia-from-hydrogen-exports.pdf

Al-Kuwari, O. (2023). *The future of liquified natural gas (LNG) in the energy transition: Options and implications for the LNG industry in a decarbonising world* [Unpublished doctoral dissertation]. University College London. https://discovery.ucl.ac.uk/id/eprint/10173650/

Anderson, K., & Bows, A. (2011). Beyond "dangerous" climate change: Emission scenarios for a new world. *Philosophical Transactions of the Royal Society A: Mathematical, Physical and Engineering Sciences, 369*(1934), 20–44. https://doi.org/10.1098/rsta.2010.0290

Andrews-Speed, P. (2015). Energy law in support of the low-carbon transition: Lessons from the United Kingdom and China. *Frontiers of Law in China, 10*(2), 295–315. https://doi.org/10.3868/s050-004-015-0016-6

Andrews-Speed, P. (2016). Applying institutional theory to the low-carbon energy transition. *Energy Research and Social Science, 13*, 216–225. https://doi.org/10.1016/j.erss.2015.12.011

Andrews-Speed, P., Zhang, S., & Wang, C. (2020). *Does 2020 mark a critical juncture in China's low-carbon energy transition?* Oxford Institute for Energy Studies. www.oxfordenergy.org/wpcms/wp-content/uploads/2020/10/Insight-76-Does-2020-mark-a-critical-juncture-in-Chinas-low-carbon-energy-transition.pdf

Bloomberg. (2020). *Hydrogen economy outlook – key messages.* BloombergNEF. https://data.bloomberglp.com/professional/sites/24/BNEF-Hydrogen-Economy-Outlook-Key-Messages-30-Mar-2020.pdf

Botzen, W., Gowdy, J., & Van den Bergh, J. (2008). Cumulative CO2 emissions: Shifting international responsibilities for climate debt. *Climate Policy, 8*(6), 569–576. https://doi.org/10.3763/cpol.2008.0539

BP (British Petroleum). (2020). *BP energy outlook 2020.* www.bp.com/content/dam/bp/business-sites/en/global/corporate/pdfs/energy-economics/energy-outlook/bp-energy-outlook-2020.pdf

BP (British Petroleum). (2021). *BP statistical review of world energy 2021.* www.bp.com/content/dam/bp/business-sites/en/global/corporate/pdfs/energy-economics/statistical-review/bp-stats-review-2021-full-report.pdf

Bradshaw, M., & Waterworth, A. (2020). China's dash for gas: Local challenges and global consequences. *Eurasian Geography and Economics, 61*(3), 266–304. https://doi.org/10.1080/15387216.2020.1716819

CarbonBrief. (2023, November 16). *China briefing 16 November: Sunnylands statement; China methane plan: Coal capacity payments.* www.carbonbrief.org/china-briefing-16-november-sunnylands-statement-china-methane-plan-coal-capacity-payments/

Chandra, V. (2017). *Fundamentals of natural gas: An international perspective.* PennWell Books.

Chapman, A. J., & Itaoka, K. (2018). Energy transition to a future low-carbon energy society in Japan's liberalizing electricity market: Precedents, policies and factors of successful transition. *Renewable and Sustainable Energy Reviews, 81*, 2019–2027. https://doi.org/10.1016/j.rser.2017.06.011

Colombo, S., El Harrak, M., & Sartori, N. (Eds.). (2016). *The future of natural gas: Markets and geopolitics.* European Energy Review. https://doi.org/10.1016/0040-1625(87)90047-3

Corbeau, A., & Ledesma, D. (Eds.). (2016). *LNG markets in transition: The great reconfiguration*. Oxford University Press for the Oxford Institute for Energy Studies and the King Abdullah Petroleum Studies and Research Center.
Downs, E. (2004). The Chinese energy security debate. *China Quarterly*, *31*(177), 21–41. https://doi.org/10.1017/S0305741004000037
Downs, E. (2021). *Green giants? China's national oil companies prepare for the energy transition*. Center on Global Energy Policy, Columbia/SIPA. www.energypolicy.columbia.edu/publications/green-giants-china-s-national-oil-companies-prepare-energy-transition/
EIA (United States Energy Information Administration). (2020). *Country analysis executive summary: China*. www.eia.gov/international/analysis/country/CHN
EIA (United States Energy Information Administration). (2023). *China country analysis brief*. www.eia.gov/international/content/analysis/countries_long/China/pdf/china-2023.pdf
EIA (United States Energy Information Administration). (2024). *Country analysis brief: Russia*. www.eia.gov/international/content/analysis/countries_long/Russia/pdf/russia.pdf
Energy Institute. (2024). *Statistical review of world energy* (73rd ed.). www.energyinst.org/__data/assets/pdf_file/0006/1542714/EI_Stats_Review_2024.pdf
ExxonMobil. (2019). *Outlook for energy: A perspective to 2040*. www.ief.org/_resources/files/events/ief-lecture-exxonmobils-2019-outlook-for-energy-a-perspective-to-2040/exxonmobil-2019-outlook-for-energy-a-perspective-to-2040.pdf
Fulwood, M., Henderson, J., Gomes, I., Lambert, M., & Sharples, J. (2020). *Emerging Asia LNG demand*. Oxford Institute for Energy Studies. www.oxfordenergy.org/wpcms/wp-content/uploads/2020/09/Emerging-Asia-LNG-demand-NG-162.pdf
Gan, Y., El-Houjeiri, H. M., Badahdah, A., Lu, Z., Cai, H., Przesmitzki, S., & Wang, M. (2020). Carbon footprint of global natural gas supplies to China. *Nature Communications*, *11*(1), 1–9. https://doi.org/10.1038/s41467-020-14606-4
Gao, G. (2015, December 10). *As smog hangs over Beijing, Chinese cite air pollution as major concern*. Pew Research Center. www.pewresearch.org/short-read/2015/12/10/as-smog-hangs-over-beijing-chinese-cite-air-pollution-as-major-concern/
Gao, Y., & Zheng, J. (2022). Clearing the air through pipes? An evaluation of the air pollution reduction effect of China's natural gas pipeline projects. *Energy Policy*, *160*(18), 112649. https://doi.org/10.1016/j.enpol.2021.112649
Gilbert, A. Q., & Sovacool, B. K. (2017). US liquefied natural gas (LNG) exports: Boom or bust for the global climate? *Energy*, *141*, 1671–1680. https://doi.org/10.1016/j.energy.2017.11.098
Gilbert, A. Q., & Sovacool, B. K. (2018). Carbon pathways in the global gas market: An attributional lifecycle assessment of the climate impacts of liquefied natural gas exports from the United States to Asia. *Energy Policy*, *120*, 635–643. https://doi.org/10.1016/j.enpol.2018.05.063
Global Energy Monitor. (2023, December). *Global gas infrastructure tracker*. Retrieved July 9, 2024, from https://globalenergymonitor.org/projects/global-gas-infrastructure-tracker/
Greenstone, M., & Schwarz, P. (2018). *Is China winning its war on pollution?* Energy Policy Institute, University of Chicago. https://aqli.epic.uchicago.edu/wp-content/uploads/2018/08/China-Report.pdf
Grübler, A. (2012). Energy transitions research: Insights and cautionary tales. *Energy Policy*, *50*, 8–16. https://doi.org/10.1016/j.enpol.2012.02.070
Guo, Y., & Hawkes, A. (2019). The impact of demand uncertainties and China-US natural gas tariff on global gas trade. *Energy*, *175*, 205–217. https://doi.org/10.1016/j.energy.2019.03.047
Huang, Y. (2024, April 24). China's battle against air pollution: An update. *Council on Foreign Relations*. www.cfr.org/blog/chinas-battle-against-air-pollution-update
ICAP (International Carbon Action Partnership). (2022). *China national ETS*. https://icapcarbonaction.com/system/files/ets_pdfs/icap-etsmap-factsheet-55.pdf

IEA (International Energy Agency). (2016). *Energy and air pollution – world energy outlook: Special report 266*. IEA Publications. https://iea.blob.core.windows.net/assets/6b75c4ae-e633-4fa0-9569-b28e226e6103/WorldEnergyOutlookSpecialReport-2016EnergyandAirPollution.pdf

IEA (International Energy Agency). (2019). *The future of fuel: The future of hydrogen*. IEA Publications. www.iea.org/reports/the-future-of-hydrogen

IEA (International Energy Agency). (2020). *World energy investment 2020*. IEA Publications. https://doi.org/10.1787/6f552938-en

IEA (International Energy Agency). (2022). *Opportunities for hydrogen production with CCUS in China*. IEA Publications. https://iea.blob.core.windows.net/assets/9c01430d-9e8f-4707-862c-35453b9e7d89/OpportunitiesforHydrogenProductionwithCCUSinChina.pdf

IEA (International Energy Agency). (2024). *Global methane tracker 2024*. IEA Publications. https://iea.blob.core.windows.net/assets/d42fc095-f706-422a-9008-6b9e4e1ee616/GlobalMethaneTracker_Documentation.pdf

IEA (International Energy Agency). (n.d.). *China*. Retrieved July 9, 2024, from www.iea.org/countries/china/natural-gas

IEEFA (Institute for Energy Economics and Financial Analysis). (2024). *Global LNG outlook 2024–2028*. https://ieefa.org/sites/default/files/2024-04/Global%20LNG%20Outlook%202024-2028_April%202024%20%28Final%29.pdf

IGU (International Gas Union). (2021). *World LNG report 2021*. www.igu.org/resources/world-lng-report-2021/

Isoaho, K., Goritz, A., & Schulz, N. (2017). *The political economy of clean energy transitions*. Routledge.

Jia, A., He, D., Wei, Y., & Li, Y. (2021). Predictions on natural gas development trend in China for the next fifteen years. *Journal of Natural Gas Geoscience*, *6*(2), 67–78. https://doi.org/10.1016/j.jnggs.2021.04.005

Jiang, H., Dong, X., Jiang, Q., & Dong, K. (2020). What drives China's natural gas consumption? Analysis of national and regional estimates. *Energy Economics*, *87*, 104744. https://doi.org/10.1016/j.eneco.2020.104744

Kaaresvirta, J., Kerola, E., & Nuutilainen, R. (2021). *How far is China from hitting its climate targets ? An overview of China's energy sector*. BOFIT Policy Briefs. https://ideas.repec.org/p/zbw/bofitb/42021.html

Khan, J., & Yardley, J. (2007, August 26). As China roars pollution reaches deadly extremes. *New York Times*. www.nytimes.com/2007/08/26/world/asia/26china.html

Koyama, K. (2020). *China's offer to achieve carbon neutrality before 2060 and its background*. Policy Commons. https://policycommons.net/artifacts/1579855/chinas-offer-to-achieve-carbon-neutrality-before-2060-and-its-background/2269625/

Le, M. T., Nguyen, T. T., Tran, V. H., Nguyen, T. K. N., & Do, H. T. (2019). Can natural gas play an important role in the energy transition in Asia in the future? *Eurasian Journal of Business and Management*, *7*(4), 28–36. https://doi.org/10.15604/ejef.2019.07.04.004

Li, F., & Li, X. (2021). An empirical analysis on regional natural gas market of China from a spatial pattern and social network perspective. *Energy*, *244*(A), 122598. https://doi.org/10.1016/j.energy.2021.122598

Li, J., She, Y., Gao, Y., Li, M., Yang, G., & Shi, Y. (2020). Natural gas industry in China: Development situation and prospect. *Natural Gas Industry B*, *7*(6), 604–613. https://doi.org/10.1016/j.ngib.2020.04.003

Li, X. (2015). *Natural gas in China: A regional analysis*. Oxford Institute for Energy Studies. www.oxfordenergy.org/wpcms/wp-content/uploads/2015/11/NG-103.pdf

Liang, C. (2023, December 21). *PetroChina looking to resume Central Asia gas pipeline line D construction in 2024: Source*. S&P Global. www.spglobal.com/commodityinsights/en/market-insights/latest-news/lng/122123-petrochina-looking-to-resume-central-asia-gas-pipeline-line-d-construction-in-2024-source

Lin, B., & Li, Z. (2020). Analysis of the natural gas demand and subsidy in China: A multi-sectoral perspective. *Energy*, *202*, 117786. https://doi.org/10.1016/j.energy.2020.117786

Liu, H., Evans, S., Zhang, Z., Song, W., & You, X. (2023, November 30). *The carbon brief profile: China*. CarbonBrief. https://interactive.carbonbrief.org/the-carbon-brief-profile-china/

Luo, G., Liu, S., Yan, X., & Guo, Y. (2021). Institutional constraints to China's low carbon transition: A case study of China's coal-to-gas program. *Structural Change and Economic Dynamics*, *57*, 121–135. https://doi.org/10.1016/j.strueco.2021.01.005

Master, F. (2024, January 17). China's population drops for second year, with record low birth rate. *Reuters*. www.reuters.com/world/china/chinas-population-drops-2nd-year-raises-long-term-growth-concerns-2024–01–17/

Mazzucato, M. (2013). Financing innovation: Creative destruction vs. destructive creation. *Industrial and Corporate Change*, *22*(4), 851–867. https://doi.org/10.1093/icc/dtt025

McGlade, C., & Ekins, P. (2014). Un-burnable oil: An examination of oil resource utilisation in a decarbonised energy system. *Energy Policy*, *64*, 102–112. https://doi.org/10.1016/j.enpol.2013.09.042

Meidan, M. (2019). *Glimpses of China's energy future*. Oxford Institute for Energy Studies. www.oxfordenergy.org/wpcms/wp-content/uploads/2019/09/Glimpses-of-Chinas-energy-future.pdf

Meidan, M. (2020). China: Climate leader and villain. In M. Hafner & S. Tagliapietra (Eds.), *The geopolitics of the global energy transition. Lecture notes in energy* (Vol. 73). Springer. https://doi.org/10.1007/978-3-030-39066-2_4

Meidan, M., Andrews-Speed, P., & Qin, Y. (2021). *Key issues for China's 14th five year plan*. Oxford Institute for Energy Studies. www.oxfordenergy.org/wpcms/wp-content/uploads/2021/03/Key-issues-for-Chinas-14th-Five-Year-Plan.pdf

Miyamoto, A., & Ishiguro, C. (2018). *The outlook for natural gas and LNG in China in the war against air pollution*. Oxford Institute for Energy Studies. www.oxfordenergy.org/wpcms/wp-content/uploads/2018/12/The-Outlook-for-Natural-Gas-and-LNG-in-China-in-the-War-against-Air-Pollution-NG139.pdf

Moe, E., & Midford, P. (Eds.). (2014). *The political economy of renewable energy and energy security: Common challenges and national responses in Japan, China and Northern Europe*. Springer.

O'Sullivan, S. (2019). *China's natural gas development report – reality check*. Oxford Institute for Energy Studies. www.oxfordenergy.org/wpcms/wp-content/uploads/2019/11/Chinas-Natural-Gas-Development-Report.pdf

Qin, Y., Höglund-Isaksson, L., Byers, E., Feng, K., Wagner, F., Peng, W., & Mauzerall, D. L. (2018). Air quality carbon – water synergies and trade-offs in China's natural gas industry. *Nature Sustainability*, *1*(9), 505–511. https://doi.org/10.1038/s41893-018-0136-7

Ren, J., Tan, S., Goodsite, M. E., Sovacool, B. K., & Dong, L. (2015). Sustainability, shale gas, and energy transition in China: Assessing barriers and prioritizing strategic measures. *Energy*, *84*, 551–562. https://doi.org/10.1016/j.energy. 2015.03.020

Rioux, B., Galkin, P., Murphy, F., Feijoo, F., Pierru, A., Malov, A., & Wu, K. (2019). The economic impact of price controls on China's natural gas supply chain. *Energy Economics*, *80*, 394–410. https://doi.org/10.1016/j.eneco.2018.12.026

Romsom, E., & McPhail, K. (2020). *WIDER working paper 2020/80. The energy transition in Asia – the role of liquefied natural gas and implications for East African producers*. United Nations University. www.wider.unu.edu/sites/default/files/Publications/Working-paper/PDF/wp2020-80.pdf

Sandalow, B. D., Losz, A., & Yan, S. (2018). *A natural gas giant awakens: China's quest for blue skies shapes global markets*. Center on Global Energy Policy, Columbia/SIPA. www.energypolicy.columbia.edu/sites/default/files/file-uploads/China%20Nat%20Gas%20Commentary_CGEP_June%202018_FINAL.pdf

Sandalow, D., Meidan, M., Andrews-Speed, P., Hove, A., Qiu, S. Y., & Downie, E. (2022). *Guide to chinese climate policy 2022*. Oxford Institute for Energy Studies. https://chineseclimatepolicy.oxfordenergy.org/wp-content/uploads/2022/11/Guide-to-Chinese-Climate-Policy-2022.pdf

Staffell, I., Scamman, D., Abad, A. V., Balcombe, P., Dodds, P. E., Ekins, P., Shah, N., & Ward, K. R. (2018). The role of hydrogen and fuel cells across the global energy system. *Energy & Environmental Science*, *2*, 463–491. https://doi.org/10.17605/OSF.IO/RZM4G

Stern, J. (2017). *Challenges to the future of gas: Unburnable or unaffordable?* Oxford Institute for Energy Studies. www.oxfordenergy.org/wpcms/wp-content/uploads/2017/12/Challenges-to-the-Future-of-Gas-unburnable-or-unaffordable-NG-125.pdf

Stern, J. (2022). *Measurement, reporting, and verification of methane emissions from natural gas and LNG trade: Creating transparent and credible frameworks*. Oxford Institute for Energy Studies. www.oxfordenergy.org/wpcms/wp-content/uploads/2022/01/Measurement-Reporting-and-Verification-of-Methane-Emissions-from-Natural-Gas-and-LNG-Trade-ET06.pdf

Stevens, H. (2023, March 1). The United States has caused the most global warming. When will China pass it? *The Washington Post*. www.washingtonpost.com/climate-environment/interactive/2023/global-warming-carbon-emissions-china-us/

Tan, N., Chang, L., & Guo, R. (2021). China's outward foreign direct investment in energy sector: The role of "intimate" relations between countries. *PLoS ONE*, *16*, 1–21. https://doi.org/10.1371/journal.pone.0254199

Tsafos, N. (2019). *How is China securing its LNG needs?* Center for Strategic and International Studies. www.csis.org/analysis/how-china-securing-its-lng-needs

UNFCCC (United Nations Framework Convention on Climate Change). (2015). *Paris Agreement under the United Nations framework convention on climate change*. Retrieved April 22, 2016, from https://unfccc.int/sites/default/files/english_paris_agreement.pdf

Vivoda, V. (2019). LNG import diversification and energy security in Asia. *Energy Policy*, *129*, 967–974. https://doi.org/10.1016/j.enpol.2019.01.073

Wang, T., Zhang, D., Ji, Q., & Shi, X. (2020). Market reforms and determinants of import natural gas prices in China. *Energy*, *196*, 117105. https://doi.org/10.1016/j.energy.2020.117105

Wei, D., Zhao, Y., Liu, H., Yang, D., Shi, K., & Sun, Y. (2022). Where will China's shale gas industry go? A scenario analysis of socio-technical transition. *Energy Strategy Reviews*, *44*, 100990. https://doi.org/10.1016/j.esr.2022.100990

Wen, Y., Cai, B., Yang, X., & Xue, Y. (2020). Quantitative analysis of China's low-carbon energy transition. *International Journal of Electrical Power and Energy Systems*, *119*, 105854. https://doi.org/10.1016/j.ijepes.2020.105854

Weng, Z., Song, Y., Cheng, C., Tong, D., Xu, M., Wang, M., & Xie, Y. (2023). Possible underestimation of the coal-fired power plants to air pollution in China. *Resources, Conservation and Recycling*, *198*, 107208. https://doi.org/10.1016/j.resconrec.2023.107208

Wood Mackenzie. (2024, May 28). *China's renewable exports grow by 35% between 2019–2023* [Press release]. www.woodmac.com/press-releases/2024-press-releases/china-renewables-overseas-expansion/#:~:text=Energy%20batteries%20surpassed%20solar%20modules,and%20Road%20projects%20in%202023

World Economic Forum. (2020). *Towards a more sustainable world: A global study of public opinion – SAP + qualtrics*. www.weforum.org/about/a-global-study-of-public-opinion-davos-2020/

World Nuclear Association. (2024, June 3). *Nuclear power in China*. https://world-nuclear.org/information-library/country-profiles/countries-a-f/china-nuclear-power

Xu, B., & Lin, B. (2019). Can expanding natural gas consumption reduce China's CO2 emissions? *Energy Economics*, *81*, 393–407. https://doi.org/10.1016/j.eneco.2019.04.012

You, X. (2022, March 29). *Analysis: What does China's coal push mean for its climate goals?* CarbonBrief. www.carbonbrief.org/analysis-what-does-chinas-coal-push-mean-for-its-climate-goals/

Yunna, W., & Yisheng, Y. (2014). The competition situation analysis of shale gas industry in China: Applying Porter's five forces and scenario model. *Renewable and Sustainable Energy Reviews*, *40*, 798–805. https://doi.org/10.1016/j.rser.2014.08.015

Zeng, B., Zhou, M., Liu, X., & Zhang, Z. (2020). Application of a new grey prediction model and grey average weakening buffer operator to forecast China's shale gas output. *Energy Reports*, *6*, 1608–1618. https://doi.org/10.1016/j.egyr. 2020.05.021

Zhang, S., & Andrews-Speed, P. (2020). State versus market in China's low-carbon energy transition: An institutional perspective. *Energy Research and Social Science*, *66*, 101503. https://doi.org/10.1016/j.erss.2020.101503

Zhao, L., Li, D., Guo, X., Xue, J., Wang, C., & Sun, W. (2021). Cooperation risk of oil and gas resources between China and the countries along the belt and road. *Energy*, *227*, 120445. https://doi.org/10.1016/j.energy.2021.120445

Zhongying, W., & Sandholt, K. (2019). Thoughts on China's energy transition outlook. *Energy Transitions*, *3*(1–2), 59–72. https://doi.org/10.1007/s41825-019-00014-w

Zhou, S., Tong, Q., Pan, X., Cao, M., Wang, H., Gao, J., & Ou, X. (2021). Research on low-carbon energy transformation of China necessary to achieve the Paris Agreement goals: A global perspective. *Energy Economics*, *95*, 105137. https://doi.org/10.1016/j.eneco.2021.105137

Zhou, Z., & Qin, Q. (2020). Decoding China's natural gas development: A critical discourse analysis of the five-year plans. *Technological Forecasting and Social Change*, *151*, 119798. https://doi.org/10.1016/j.techfore.2019.119798

Zou, C., Zhao, Q., Chen, J., Li, J., Yang, Z., Sun, Q., Lu, J., & Zhang, G. (2018). Natural gas in China: Development trend and strategic forecast. *Natural Gas Industry B*, *5*(4), 380–390. https://doi.org/10.1016/j.ngib.2018.04.010

6 The LNG industry in the future

This chapter provides an analysis of the LNG industry in four parts. The chapter starts by considering the three country cases presented in Chapters 3–5. Based on the individual analyses from those chapters, I discuss similarities and differences between the energy transitions and predicted future of LNG in the UK, Japan, and China. Grubb et al.'s (2015) "Three Domains" framework is introduced and used to understand three pathways through the energy transition, as exemplified by the three country cases. The second part of the chapter is an overview of the threats and opportunities facing the LNG industry. Here I adopt a broader global perspective drawing on in-depth interviews with 17 LNG industry participants (based on Al-Kuwari, 2023). The third section of the chapter is a technical analysis of the industry, addressing two main questions: What measures can the LNG industry take to address decarbonisation, and what is the probable impact of hydrogen fuel on the industry? The chapter ends with a short summary and T-E-T-I analysis (as introduced in Chapter 2) for the LNG industry as a whole.

Integration of the UK, Japan, and China case study findings

The UK and Japan have been gas users for decades, yet demand has generally peaked and plateaued, while Chinese demand continues to grow at high rates. In all three countries, significant gas infrastructure has been built to reflect the importance of gas in energy systems. The UK is historically a major producer of natural gas, but it has limited reserves. China is an important gas producer with significant reserves. Japan has historically not been a gas producer and has negligible reserves. The UK and China import both LNG and natural gas, while domestic production supports the development of pipeline infrastructure. Conversely, Japan is fully dependent on LNG for all gas imports, which have historically been oil indexed on a long-term basis. A new price marker, the Platts Japan Korea Market – also known as the "spot market" – is gaining influence, especially in the short term. The UK's National Balancing Point is a virtual trading hub used as the basis for trading gas and for LNG imports in the UK. The Chinese LNG market uses a mix of contracts, with some price exchanges emerging. All three countries have decarbonisation pledges. Japan and the UK have legally binding commitments to achieve net-zero by 2050, while China aims for carbon neutrality by 2060. All three countries have

DOI: 10.4324/9781003545149-6

Case	Gas demand / 2006 – 2016 growth	Indigenous production / reserves	LNG imports	Pipeline gas imports	Gas-on-gas competition	Decarbonisation pledges	Political system
Japan	117 BCM + 2.9 %	X (3 BCM) N/A	113.9 BCM 16 terminals	X No pipeline gas connections	N/A	• • Hydrogen economy • 80% reduction by 2050 (base year not specified)	Multi-party bicameral parliamentary representative democratic constitutional monarchy
UK	78 BCM –2.4 %	• • 42 BCM • 6.5 TCF (0.1% share of total)	7.2 BCM 3 terminals	39.4 BCM Interconnector and BBL	Liquid Market	• • EU – 80-95% 1990 by 2050 • UK – net zero by 2050	Parliamentary democracy under a constitutional monarchy
China	240 BCM 13.7 %	• • 149 BCM • 193 TCF (2.8% share of total)	39.4 BCM 12 terminals	52.6 BCM West to East gas pipeline	Mix of contracts	X • Peak emissions by 2030 • Focus on air quality and carbon intensity	Socialist republic run by the Communist Party of China

Figure 6.1 Summary of key characteristics for Japan, the UK, and China (developed by Al-Kuwari, 2020 based on BP Statistical Review of World Energy, 2021)

different political systems that affect how energy and environmental policies are set and implemented. Figure 6.1 presents a visual summary of the three country cases discussed in the preceding chapters, where the countries are contrasted in terms of their gas demand, gas reserves, LNG imports, decarbonisation pledges, and political systems.

Table 6.1 provides a summary of the country cases using the T-E-T-I framework (introduced in Chapter 2). Through its application, three interconnected themes were identified for determining the role of natural gas use and LNG in the energy transition: (1) energy security, (2) climate change and environmental policies, and (3) affordability and competitiveness. These three themes mirror the "energy trilemma" identified by the World Energy Council (2021, p. 2) as the balancing of three main goals – energy security, energy equity, and environmental sustainability. Energy security and security of supply relate to countries being able to meet energy demand through developing a resilient infrastructure that can sustain system disruptions and shocks (World Energy Council, 2021). In the electricity area, the trilemma is less relevant because the structure is related to boundaries rather than trade-off between environment and competitiveness, and less in conflict with energy security. In energy markets using mainly fossil fuels, including LNG, a significant trilemma still exists, with the cost for decarbonisation adding another dimension. I now discuss the three themes that affect how institutional, social, and resource characteristics affect LNG and natural gas use in the UK, Japan, and China.

Energy security

The role of LNG in the context of energy security and decarbonisation depends on a country's domestic energy supply alternatives and the pace of decarbonisation

Table 6.1 Case study summaries using the T-E-T-I and transition context frameworks (developed by Al-Kuwari, 2022)

Country	*Resources*	*Market*	*ET dynamics*	*Power*	*Transition context*	*Key themes*	*Role of LNG in ET*
United Kingdom	Developed gas and LNG infrastructure; EU interconnection; decreasing domestic production	Competition focused; fully liberalised	Leading by example; COP26; hard-to-abate sectors will be more challenging (i.e. heating)	Brexit; alignment around decarbonisation agenda	Purposive transition	Climate change and leading by example; decarbonisation; North Sea, CCS, and hydrogen; energy security; competitiveness	Seasonal transition fuel to electrification and hydrogen; backup transition fuel of choice
Japan	Developed LNG infrastructure; no pipeline connection; no domestic production; nuclear future main variable; limited renewables growth	Gas reforms; energy security – 3S + 1	1997 Kyoto Protocol; Japan 2050; hydrogen economy; anti-nuclear sentiment	Iron triangle, politics; bureaucracy; industry	Purposive transition	Nuclear energy and Fukushima, energy security; the iron triangle; competitiveness and the energy mix; climate change and GHGs	Decarbonisation ambitions with short- to medium-term options
China	LNG infrastructure developing rapidly; domestic coal supply; misalignment supply-and-demand regions	Moving towards liberalisation; gas demand still growing; Pipeline versus LNG	Driven by air pollution; dual carbon goals 2030/2060	Top-down, centralised; BRI; central versus regional	Endogenous renewal	The role of coal; energy security; climate change politics and commitments; central versus regional dynamics; vested interests; market liberalisation	"All of the above"; environmentally superior to coal but must be affordable; meet growing demand

efforts. In the UK, decarbonisation goals are advancing energy security measures. When the UK switched from being an exporter to an importer of fossil fuels, coal gradually became less desirable (despite remaining domestic reserves), and LNG and imported pipeline gas provided a way to leverage developed gas infrastructure while reducing emissions supported by the "dash for gas". LNG provides a lower carbon, stable and predictable, backup option for seasonality and heating. In contrast, by managing a shortfall in nuclear production, LNG proves reliable in supporting Japan's energy security agenda. Environmentally, LNG is more attractive than coal and other fossil fuels (mainly when it comes to electricity). LNG can also play a role as a potential bridge fuel to a hydrogen economy. In China, energy security currently conflicts with the ability to meet decarbonisation targets – the supply of domestic low-carbon resources is outpaced by growing demand. LNG supports meeting coal-to-gas switching demands given limits on domestic gas production and pipeline imports. However, increased imports would still need to compete with domestic coal, which could negatively affect China's energy security goals despite reducing overall emissions.

Increased dependence on LNG imports is considered politically unfavourable from an energy security perspective. LNG is perceived as contributing to trade imbalances, creating geopolitical risk, and increasing exposure to market volatility and supply dependability. In Japan, nuclear energy is the key variable. Here, and in most cases, nuclear is a domestic resource, developed to support a country's energy security agenda. LNG imports place a major burden on Japan's finances. The "Iron Triangle" vested interest dynamic in Japan favours maintaining a balanced role for all fuels, increasing the importance of nuclear and coal. In China, there is concern about the availability of competitively priced LNG to meet the country's growing demand, especially when compared with domestic coal, and hesitance to over-rely on imports, especially when domestic gas reserves are stagnant. LNG imports can be made more secure through policy measures and commercial arrangements (domestic and international). The UK has been a leader in developing efficient gas markets with open access to LNG. China and Japan engage in resource diplomacy, invest in upstream LNG projects, and diversify import sources. An increase in LNG supplies coupled with an increase in suppliers and market participants may render LNG more reliable in the eyes of policymakers and energy planners.

Climate change and environmental policies

Climate change policies encourage decarbonisation and support economic development. International cooperation around the Paris Agreement and the net-zero agenda has created opportunities for exporting knowledge, skills, and technologies. This is, however, to the long-term detriment of fossil fuel use, including LNG imports.

The UK took a "lead by example" position and pursued aggressive domestic targets early on, supported by the EU's decisive move towards net-zero. Japan is associated with the 1997 Kyoto Protocol and has cumulative emissions responsibilities towards developing nations, placing the country under pressure to lead R&D and technology deployment to mitigate the impact of climate change. However, Japan does not experience regional pressure and is not tied to a regional framework. Its domestic needs will thus likely play the determining role in any measures it takes. China is key to the global climate progress because it will be the world's leading cumulative contributor of CO_2 by 2050. The country's global standing is slowly moving away from being a "climate villain", supported by its announcement of dual 2030/2060 carbon goals.

Decarbonisation and net-zero ambitions shape the role LNG can play. Natural gas and LNG have not carved out a niche as a clean fuel, separate from other fossil fuels. The EU leads the way on regulations affecting the future of LNG, despite (and possibly because of) stagnating long-term gas demand. The European green deal offers some extension for LNG use, but the industry has generally failed to strengthen the environmental case for LNG as a pragmatic solution over the next 10–15 years. LNG exporters will have to adapt to EU rules related to carbon measurements and potential carbon border adjustment mechanisms that could affect imports later this decade. Increased scrutiny of CH_4 emissions by policymakers and NGOs will impact LNG exporters to varying degrees. In Japan, environmental awareness is a pillar of the 3 E+S policy framework. There, LNG is still part of the solution when it comes to meeting climate objectives. In China, climate objectives linked to air quality and GHG reductions have become more important. The impact of environmental degradation on the population makes the environment a priority in energy policy, suggesting that LNG may be perceived as a cleaner alternative to coal.

Affordability and competitiveness

Affordability is a key, yet unequal, policy driver in all three countries. Market liberalisation was a major objective for UK policymakers in the late 1990s and early 2000s. As the energy system moved to become import-based, a market-based policy paradigm developed. Policymakers viewed an efficient energy market as the most suitable mechanism for managing energy supply as a tradeable commodity. Market policies spurred investment into the gas market, which led to "gas-on-gas" competition (underpinned by a liquid gas market). In Japan, supply security and reliability were deeply embedded in the development of their energy structure, taking priority over other factors. Japanese import prices have consistently been among the highest globally. Nonetheless, large utilities (including partial SOEs) can transfer the procurement costs of imported fuels to consumers, reducing the incentive for importers to be globally competitive. However, this supply certainty–focused structure led to a disconnect from international markets, low competition, and limited ability to support major shifts in the energy sector. More recently, competition has been linked to energy security to ensure access to supplies, with physical supply security remaining the central focus. Uncertainty regarding a return to nuclear power and limited potential for growth in renewable power emphasises LNG's central role in Japan's energy security over coming decades. Chinese policymakers' focus on air-quality goals has led to an increase in gas consumption. Gas demand spurred market-oriented price reforms aimed at reducing the overall cost to gas consumers. The liberalisation of the LNG market ties into the Government's objective to have a more flexible and competitively priced gas market. Market liberalisation encourages LNG imports, where third-party access can unlock demand for downstream and wholesale gas users. Previous efforts at government control of natural gas prices led to price distortions and cross-subsidies. Linking LNG prices to the global market allows for more market players to meet growing LNG and pipeline gas demand. The long-term role of LNG in meeting increased Chinese energy demand will largely depend on its ability to be a comparatively affordable fuel source.

The potential economic impact of decarbonisation policies on consumers is particularly challenging in the UK and Japan. The UK government will find it difficult to convince consumers to accept additional costs for a "green agenda" without clear benefits, especially after the COVID-19 pandemic and Brexit. Regulators must manage gas policy while promoting decarbonisation and investment into new technologies and infrastructures. In Japan, utilities have the dual role of supplying affordable energy while working towards decarbonisation. Increased economic pressure could reduce revenues and the ability of companies to invest in infrastructure that supports the energy transition. Natural gas is, however, well established in both the UK and Japan. A wide range of consumers (from large utilities to homeowners) are accustomed to using and paying for it. In certain UK sectors (such as heating), natural gas will have strong cost advantages when retrofitting costs are taken into consideration. In the short term, carbon trading may help the gas industry because it supports converting coal to natural gas for heating, utilities, or as feedstock.

Coal is unlikely to play a future role in the UK energy system given competitive renewables and the decarbonisation agenda. However, the low cost of coal makes it an option in Japan and China. In Japan, relatively high LNG prices might allow coal to maintain some role, and without continued support for coal-to-gas switching via a carbon tax or other measures, coal could become a viable option for importers. LNG might be viewed as ineffective for balancing economic and environmental goals in a non-competitive environment. In China, coal continues to play a role in both energy security and economic development pillars because of its domestic availability and strong political base. Non-competitive LNG prices and the lack of aggressive gas promotion signals could encourage funding for "clean coal" (or coal use in general).

Three domains and three pathways

Grubb et al. (2015) developed the Three Domains framework to demonstrate that energy systems develop around policies derived from three domains:

> Domain 1: Satisfice – is focused on the present and short term. Based on principles of behavioural and organisational economics, decisions from this perspective are about being "smarter".

> Domain 2: Optimise – centres on the medium term. It is based on neoclassical theories where trade-offs substitute for cleaner and optimised products.
>
> Domain 3: Transform – is focused on long-term transformational change following evolutionary and institutional theories. Through strategic decision-making, the current best practice frontier advances via innovation and the evolution of complex systems.
>
> (Grubb et al., 2015)

The framework is useful for assessing the future impact and implications of the LNG industry. Figure 6.2 shows the three domains and the three policy pillars (standards and engagement, markets and prices, and strategic investment) mapped against the short-, medium-, and long-term options for the industry.

In domain 1 (column 1, Figure 6.2) the LNG industry has established best practice environmental, commercial, and safety standards and regulations along the value chain. The industry has become more economically and operationally efficient by developing technologies like flare reduction, reliquefication on board vessels, and standardising certain clauses in contracts. In this domain, the industry is moving towards measuring, managing, and reducing vented and flared emissions. LNG vented CH_4 and upstream emissions were mostly overlooked in the past because they were not considered important and/or because the industry needed better systems in place to deal with them. However, major industries are now paying closer attention to operations that lead to gas wastage and unwanted emissions. In domain 2 (column 2, Figure 6.2), the industry moves towards optimising the entire value chain as it is today. The goal in this domain is to move towards cleaner products and processes, and ultimately reduce the carbon footprint of LNG, specifically at the point of combustion. Carbon-neutral cargoes and CCS tools are means to reduce LNG-related emissions in this domain. In domain 3 (column 2,

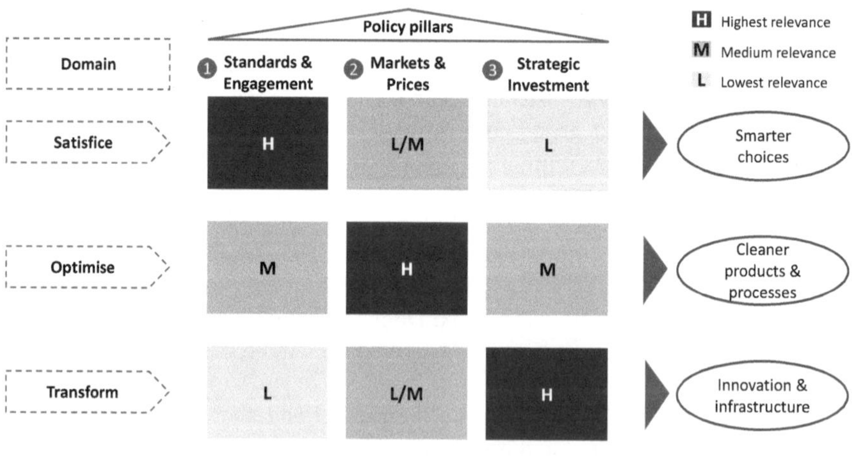

Figure 6.2 Three domains of sustainable development (Grubb et al., 2015)

Figure 6.2), the LNG industry explores strategic investments in new directions to transform the industry so that it is compatible with the net-zero ideal. Multinationals and NOCs approach decisions in a long-term and strategic manner. Smaller players – including traders and those participating in only one part of the value chain – are confined by existing market structure and trends. Strategic actors must consider timelines beyond several decades. Hydrogen and its associated carriers represent a potential pathway via a landmark change. This can transform LNG into a low-carbon option but requires strategic investment and radical innovation. Here the value chain is transformed using available tools including CCS, and blue hydrogen, although there is an element of uncertainty.

Figure 6.3 maps the three country cases against the three domains framework. The cases highlight three distinct LNG pathway types based on the analyses of the preceding chapters. The UK is seen to follow a transformation pathway where energy security and climate change are linked. In this pathway, LNG plays the role of a seasonal and transitional fuel to support aggressive electrification and the transition to a zero-carbon energy system. Japan appears to be following a transition fuel pathway where energy security is the prominent policy goal. China's path is one of integrated development, a pathway that straddles all three domains with a focus on affordability and energy security. The three countries are at distinct stages of their energy system development, which is reflected in their energy policies and strategies. China, focused on energy intensity and air quality, is mostly optimising in the second domain while Japan, with its "Hydrogen 2050" is planning on transforming its energy system through new infrastructure in the long term. At this stage, the security of lower carbon supplied to meet industrial and power generation needs still drives policy. The UK is moving past the optimising pillar towards transformation of the energy system, as shown by its recent focus

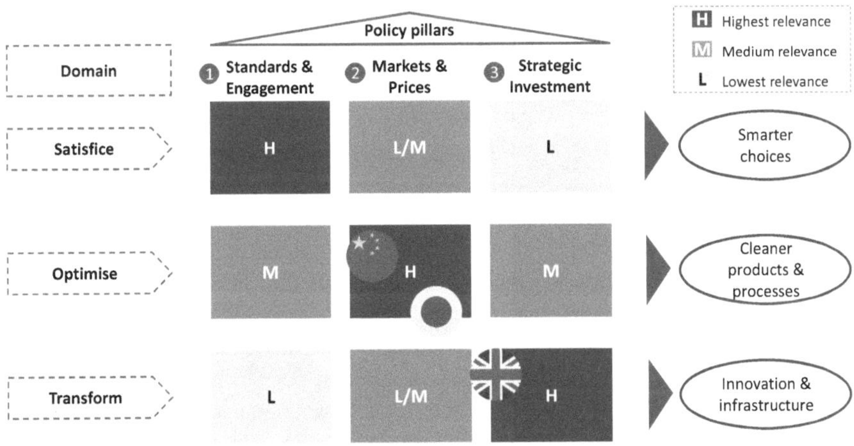

Figure 6.3 Case study selections mapped onto the three domains of the sustainable development framework (adapted from Grubb et al., 2015)

on renewable energy development and energy efficiency, and its current focus on decarbonisation.

UK: Transformation pathway

The UK places great emphasis on its net-zero targets, with climate change commitments and energy security being key priorities. Policymakers are focused on the development of industrial clusters and offshore wind. These will underpin hydrogen development, both to decarbonise the energy system and to reduce energy dependence. The decarbonisation of heating and the role of natural gas therein will have a major impact on LNG's role as a reliable seasonable backup option. Continued efforts to lower value chain emissions and increase overall emission transparency will also play a crucial role. The UK has displayed global leadership in setting decarbonisation targets. Its political and socio-economic system has high adaptive capacity, and a history of efficient policymaking and a strong energy and research industry. The country is likely to undergo a purposive transition with rapid decarbonisation occurring in the coming decades. Figure 6.4 illustrates the evolution of the three energy trilemma priorities – energy security, climate change and environmental policies, and affordability and competitiveness for the UK. As for all countries, the balance between these three priorities reflects national circumstances, natural history, and events in the context of the energy transition. There is an interplay between the themes, and if policymakers focus too much on one of them, it will be to the detriment of the others.

Japan: Transition fuel pathway

In Japan, there is a relative balance of the three themes with priority placed on energy security. Japan is reliant on imports and has established a robust and balanced system of fossil fuels and import sources. The future of hydrogen is a major focus and there is significant R&D investment in the country. LNG has an opportunity to serve as a "bridge fuel" and uncertainty around nuclear and limitations around renewables may mean that LNG will remain key to supply security. The main pathways will be to provide stable and reliable supplies while taking affordability and GHG emissions of the value chain into consideration. Affordability will increasingly overlap with energy security because of exposure to volatile LNG prices, especially at times of supply tightness. Higher levels of coordination between government and industry, and a willingness to transform to a hydrogen economy may position Japan on the right transition track. Path dependency related to the "iron triangle" and other barriers position the energy system on a pathway in which LNG will be a fundamental transition fuel. Japan is following a transition pathway where LNG supports its energy security priorities to cover nuclear electricity shortfalls, and in the longer term supports the shift towards a hydrogen-based economy. Japan's path through the energy transition is graphically depicted in Figure 6.5.

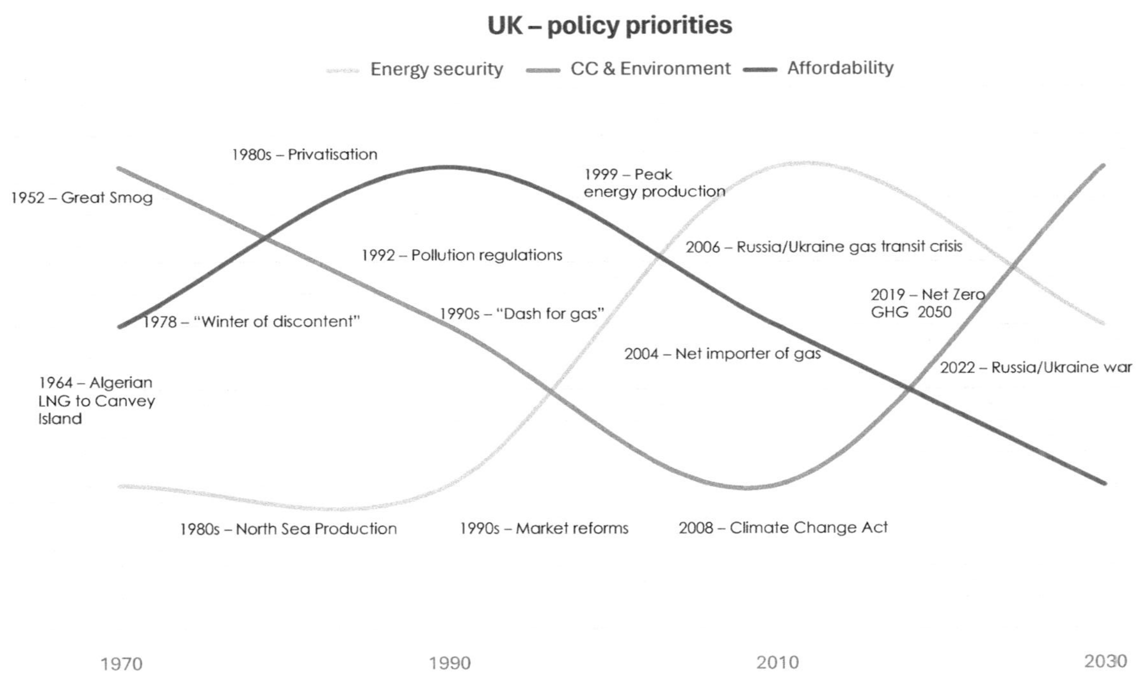

Figure 6.4 Evolving UK energy-related policy priorities 1970–2030 (developed by Al-Kuwari, 2022)

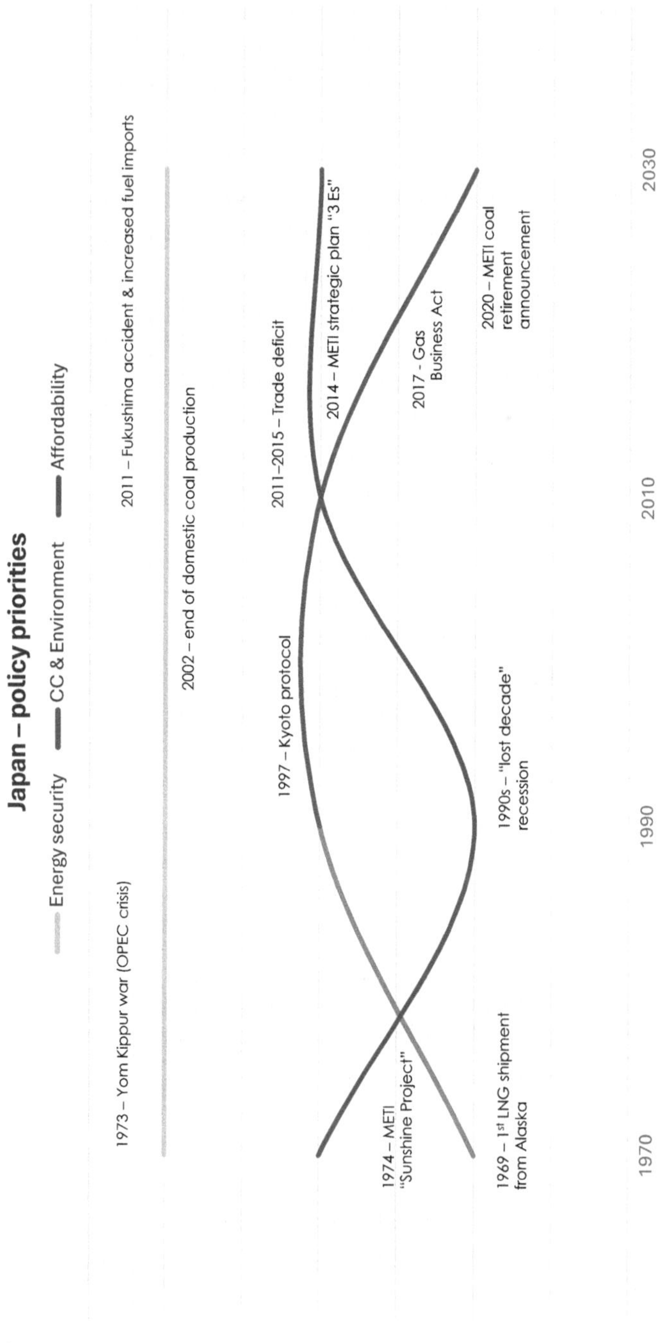

Figure 6.5 Evolving Japanese energy-related policy priorities 1970–2030 (developed by Al-Kuwari, 2022)

China: Integrated development pathway

China's priority is to meet its rapidly growing energy demand in an affordable manner and at a high level of domestic self-sufficiency. Along with domestic resources and access to pipelines, the 2060 target allows Chinese policymakers to take a pragmatic approach to imports. Competitiveness is also key to continued growth. The Chinese economy must continue to grow if its importers are to afford volatile and sometimes high-priced LNG. Air quality and access to more supplies will also be determining factors. Carbon pricing, technological advances, emission reductions, increased global supplies, and a liberalised market could favour LNG. However, China's growth trajectory does not place it on a "development-first" pathway. The country's development straddles all three domains. China is pursuing approaches characteristic of domain 1. Yet, the role of air pollution and the significant investment in green technology and innovation suggest an "all of the above" approach. The country's significant size, regional characteristics, and rapid economic development strategy all point to an integrated development pathway (Figure 6.6).

Impact of the energy crisis

Affordability and security of supply have moved up in priority on policymakers' agendas during the energy crisis. For Europe and the UK, the crisis has increased concerns about gas dependence. The UK is committed to decarbonisation and has linked this to energy security. The energy crisis is likely to shorten the gas bridge and accelerate energy transformation. In Japan, the energy crisis has highlighted the importance of nuclear energy and supported policymakers seeking to reintegrate nuclear power to mitigate energy security concerns. In China, where the COVID-19 pandemic suppressed LNG demand, domestic coal will help reduce high LNG prices when needed. In the medium term, both China and Japan could have access to excess LNG supplies entering the market in coming years. This will likely be spurred by additional investments (e.g. see Ouki, 2023 for discussion of the potential African gas "renaissance" and investments made by China and Japan in the African market) and reduced long-term gas demand in Europe. As supply and demand balance, affordability and energy security could do so as well. Figure 6.7 demonstrates the country pathways of the case studies, and the different LNG industry domains from 2020 to 2060.

In summary, development-first pathways – energy systems that are primarily driven to meet increased and more affordable demand – will continue to be aligned with domain 1. This category covers new price-sensitive and air pollution-driven LNG markets such as Bangladesh, the Philippines, and Pakistan, which appear set to prioritise meeting affordable demand for several decades. LNG in integrated development pathways, such as China, will continue to operate in satisfice and optimise roles for the next several decades too, with affordability as a primary driver balanced alongside energy security and environmental concerns. LNG in transition fuel pathways will play an optimising role for the next decade, eventually transitioning to transformation roles. In Japan and the UK, LNG already plays that

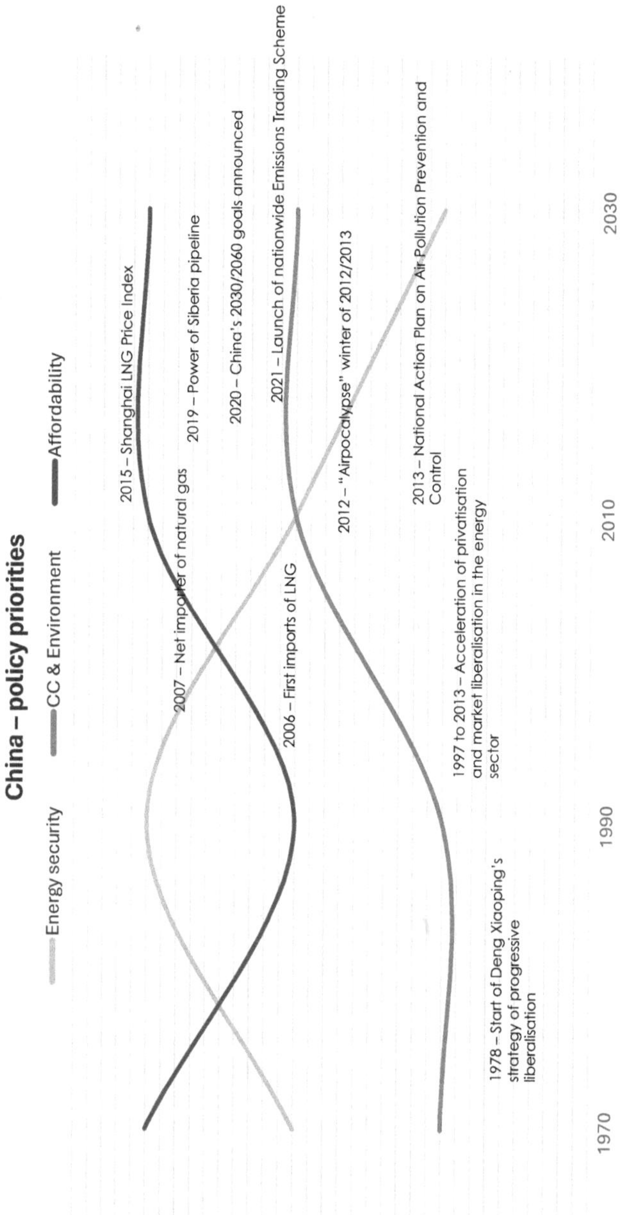

Figure 6.6 Evolving China energy-related policy priorities 1970–2030 (developed by Al-Kuwari, 2022)

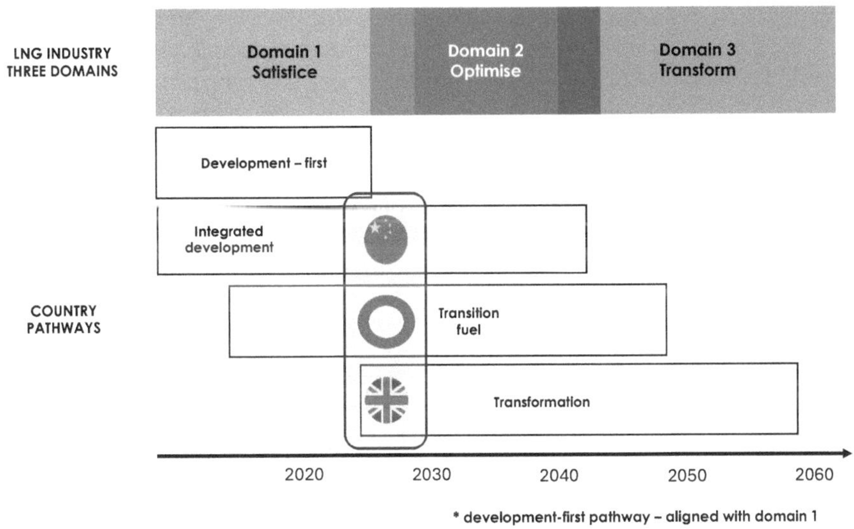

Figure 6.7 Illustrative LNG industry domains and country pathways over time

role, and policies and investments towards more radical decarbonisation solutions are already noted. Ultimately, in a post-2060 net-zero world, LNG will have to increasingly operate in this space as systems evolve.

Energy-transition dynamics

The challenges and uncertainties presented by the potential impact of decarbonisation measures in key markets increase the risk for investors. This is especially the case for greenfield projects where both exporting and importing countries impose regulations (e.g. CCS for Australia). The future role of LNG in energy systems is inextricably linked to natural gas. Forecasts project peaking circa 2030 to 2045, depending on the decarbonisation scenario. LNG can drive coal-to-gas switching and to some extent oil switching subject to infrastructure development and overall affordability. However, in developed markets, increased decarbonisation measures and increasingly competitive renewables and battery technologies are already pushing gas out in certain power markets. In the long term, green hydrogen could prove to be a direct competitor. Natural gas-based hydrogen presents an opportunity to support LNG demand in the short to medium term. The LNG industry will need to find ways to compete with coal and renewables for both price and carbon footprint. Affordability of LNG will be linked to profitability of LNG projects in the face of increased emissions scrutiny. Pipeline gas could limit LNG expansion in major potential markets such as China.

Countries assessing options for their fuel mix must consider many variables. Conditions for natural gas to be a successful transition fuel include timing, climate ambition, CH_4 emissions, substitution, carbon lock-in, and the availability of CCS technology. Conditions related to technical and economic improvements in renewables may dictate that some countries bypass gas consumption altogether, especially for power generation purposes, focusing on low-carbon sources such as wind and solar. Breakthroughs in electricity storage and affordability, compounded by the lack of gas infrastructure, could accelerate this phenomenon or see a peak of consumption within a few decades. Limited funding and the lack of natural gas infrastructure are key constraints for many African countries (Energy Capital & Power, 2024). While North African countries such as Algeria and Egypt have extensive natural gas transmission and distribution networks, most of the other gas-producing countries have limited gas pipeline transportation capacity. Nigeria's "Natural Gas Master Plan" in the early 2000s aimed to develop a national gas infrastructure but most of its targets have yet to be achieved. These shortcomings reflect policy and regulatory constraints and lack of funding (Energy Capital & Power, 2024). Furthermore, many of the African domestic gas markets are small with limited commercial incentives for expanding infrastructure beyond existing gas supply project boundaries or nearby off-takers.

An increased focus on hydrogen, given its flexibility and ability to target hard-to-abate sectors along with improvements in electrolysis, indicates a significant role for hydrogen within a few decades. Fuel-switching for air-quality purposes, along with increased industrial use, will continue to drive natural gas growth in the short to medium term. However, uncertainties persist in the long term. Natural gas exporters must consider how factors such as the evolution of synthetic gases and electrolysis-produced hydrogen, continued electrification through renewable energy sources, and potential breakthroughs in battery technology could affect their global market share. The Gas Exporting Countries Forum (GECF) comprises 19 countries[1] that are leaders in natural gas production. GECF (2023) data suggest that policymakers have "never faced such huge challenges [as in the present] in solving the energy trilemma of achieving security, affordability and sustainability" (p. 5). Together, the GECF member countries play a central role in the global gas trade, accounting for 47% of the world's pipeline gas exports and 52% of global LNG exports (GECF, 2023). The forum provides a platform for reinforcing global energy ties and managing the many challenges of the energy transition in a collaborative way, promoting cooperation and capacity development needed to maximise the full potential of LNG.

Opportunities and challenges for the LNG industry: findings from stakeholder interviews

This section provides insights into opportunities and challenges for the LNG industry identified by a global pool of LNG industry experts and stakeholders. In-depth individual interviews were conducted with these experts by Al-Kuwari (2023) (and see Al-Kuwari, 2023 for further details of the selection and recruitment of the interviewees, interview transcripts, and more detailed analysis).

Decarbonisation is viewed as both an opportunity and a threat for the LNG industry. On the one hand, the structure of the value chain and technological and operational improvements support opportunities for increased transparency for emission disclosures and reducing overall GHG. On the other hand, increased decarbonisation targets and further delays in the industry regarding CO_2 and CH_4 emissions could make LNG increasingly unattractive for potential importers. Interviewees determined that upstream regulation, technological improvement, carbon prices, and societal and investment pressure will largely determine decarbonisation pathways for the LNG industry. Uncertainty around the future of the LNG industry increases with the speed of the global decarbonisation agenda. The need for improved air quality will remain a key driver of gas demand in the developing world. However, there is also a longer-term requirement to demonstrate how gas can contribute to global decarbonisation efforts in a net-zero world.

LNG is a flexible solution that can play several important roles in the energy transition. Gas represents a "lower carbon route to a low carbon future" (Al-Kuwari, 2023). LNG is uniquely positioned to meet a wide range of needs within and beyond the transition fuel narrative. In advanced economies, it may leverage established gas-fired generation and maintain stability and resilience in energy systems. This could occur as the use of less predictable renewable energy generation methods increases. Moreover, the possible phasing out of nuclear power could create an opportunity to provide stable baseload-like supply. There is a clear role for LNG in fast-growing economies that could meet demand growth over the next few decades before more decisive moves towards net-zero. Interviewees identified two main opportunities for the industry: (1) coal-to-gas switching, and (2) emerging gas markets and sectors. Both opportunities can be supported by improved economics and technology.

In Asian countries, LNG is well positioned to play a growing role under the "all of the above" approach that economies will pursue. China is key to the future of LNG. Here, demand will depend on economic conditions, coal use, pipeline imports, domestic gas production, and government support for increased natural gas consumption. Several 2°C scenarios foresee an increase in natural gas consumption because it represents the most efficient way to meet rapidly rising Asian demand. In the IEA's Stated Policies Scenario (STEPS), around half of global gas demand growth comes from developing Asian economies with China alone representing about one-third of that growth between 2020 and 2030 (IEA, 2021b, 2023b). Natural gas currently meets around 12% of primary consumption in the Asia-Pacific region (IEA, n.d.). Increasing that to 20% would add the equivalent of 400 million tonnes of LNG to annual gas demand, doubling LNG's current market size (Bresciani et al., 2020). In both India and China, natural gas has a larger role in Rapid Transition scenarios due to increased coal-to-gas switching in the medium term to meet increased demand (BP, 2020). For many stakeholders, LNG is the best option for decarbonising Asia. Hydro, nuclear, and renewables will not be able to satisfy electricity and industry demands. Interviewees expressed more certainty about the increased role of LNG in China compared with India. A lack of infrastructure and questions around affordability are potentially major obstacles for Indian LNG development. Demand from emerging Asian countries such as Thailand, Vietnam, Singapore, Pakistan, the Philippines, Bangladesh, Malaysia, and Indonesia could be significant.

Coal displacement presents a key opportunity for the industry over the next 20–30 years. Moving from coal and oil products to natural gas is often driven by market-based pricing, influenced by (1) carbon pricing, (2) policy measures to phase in certain fuels or applications (such as boiler conversion programmes), and (3) attempts to de-bottleneck physical supply chains. In mature systems, gas production can replace older less efficient coal plants. In the medium term, power generation is responsible for half of gas substitution based on coal reduction, development of domestic natural gas, and a reduction in oil imports. The case for coal switching is easy to justify in India and China. Similar possibilities exist for switching away from liquid fuels, diesel, and heavy fuel oils in other regions, including the Caribbean, Latin America, and West Africa. Here, LNG can support bridging to domestic gas development or substitute other fuels, assuming there are supportive government frameworks to attract significant financing for LNG projects. Coal accounts for 47% of Asia's and 23% of Africa's power consumption (Bresciani et al., 2020; IEA, n.d.), suggesting that switching is a viable option. However, infrastructure could be a bottleneck for new markets. In Asia, for example, LNG regasification capacity needs to double every five years for LNG to meet predicted demand growth (Fulwood et al., 2020), and in Africa natural gas infrastructure remains a key constraint for many current and potential gas-producing countries (Ouki, 2023).

In Europe, the coal-to-gas narrative has less value because most of the switching has already taken place. Nonetheless there are still pockets of demand (notably in Eastern Europe) where natural gas offers a practical and cost-effective way to reduce carbon intensity through leveraging established gas infrastructure or "easy to bring on" LNG. LNG's "plug and play" type solution is viewed as achievable, implementable, and scalable, attractive to governments looking to expand their energy system at scale, especially if LNG is affordable. The energy crisis related to the Russia/Ukraine war forced countries to revisit their approach to coal to mitigate against Russian gas shortages and meet domestic demand.

Flexibility around LNG import technology and power solutions minimises the lock-in risks of other alternatives, including complex pipeline gas projects. LNG import terminals can provide additional flexibility where needed, and FSRUs can quickly respond to shortages. This can be supported at the other end of the chain by technology that encourages monetising previously uneconomical reserves. Interviewees also noted the potential to use LNG for heavy-duty vehicles and shipping. The industrial sector – anticipated to be the main sector driving demand in the medium term – will be responsible for 45% of gas consumption growth from 2023 to 2024 (IEA, 2024b), and was highlighted as a key target for future supplies. Table 6.2 provides a summary of the three main challenges for the LNG industry identified by interviewees (Al-Kuwari, 2023) along with potential mitigating factors and opportunities.

Table 6.2 LNG industry participants interview analysis (Al-Kuwari, 2023)

Challenges	*Issues*	*Mitigating factors and potential opportunities*
Technology and alternatives	Emerging renewables with batteries Green hydrogen	Backup in developed markets Address seasonality Competitive economics
Rapid transition	Skipping the gas bridge Stranded assets Climate change policies	Coal to gas switching Established infrastructure Decarbonising the value chain Time-horizon
Evolving market dynamics	Financing requirements Market conditions Societal pressure/public opinion	Overall gas supply Flexibility Improved lobbying and communication

Renewables with batteries

Battery storage combined with renewables is the biggest threat to the future of LNG in the energy transition. Increased efficiency of battery storage has played a major role in underpinning growth of investment in utility-scale renewable power. In 2023, global investment in battery storage exceeded US$35 billion, contributing to the deployment of over 85 GW of global installed battery storage capacity, which was reflected in a continued decrease in battery costs (IEA, 2023b, 2023c, 2024a). Renewables combined with battery storage are projected to have a direct negative impact on gas demand, specifically gas-fired power generation, largely driven by continued cost reductions. Renewable energy is being integrated more effectively in the power sector, aided by improved weather pattern forecasting, big data analytics, machine learning, and battery technology.

As battery costs decline and renewable power becomes less intermittent, the value of LNG as a peak generator will decrease. Interviewees agreed that LNG will initially be most impacted in Europe because of aggressive decarbonisation targets and expected sustained support for renewable and battery technology. However, Japan, South Korea, and other countries with more limited renewable resources will continue to depend on LNG until (1) LNG is no longer considered the best option to "back fill" renewables (unlikely to occur before 2035), and (2) when battery technology or green hydrogen become cost-competitive at scale. In emerging gas countries like India, the role of battery technology will partly depend on the extent of investment into gas infrastructure over coming decades. LNG could play the role of a facilitator. Depending on the market, LNG has distinct advantages over underdeveloped battery technology in the short to medium term. In Europe, renewables are rising on the demand side and flexible indigenous production is declining on the supply side. Additional flexibility is needed to address major demand swings, and a strength of the LNG value chain is its ability to manage weekly and monthly demand. Additional import terminals can be built, and floating regasification terminals can be deployed at relatively short notice, thus providing significant send-out flexibility. Ouki (2023) describes the potential of sub-Saharan African countries (e.g. Mozambique) to meet the LNG needs of European and Asian countries from 2030 as new LNG sources come onstream. However, he cautions that flexible contractual frameworks will be needed to remain competitive with the United States and other established markets.

Several interviewees cited the 2018 European winter and the 2021 Texas power crisis as examples of the positive role of gas. In a system with less coal and nuclear, gas may be the best option to manage such scenarios, highlighting the potential long-term strategic value of LNG in a rapidly decarbonising seasonal market like the UK. LNG could play a similar role to that of hydropower in Norway, that is, large flexible volumes to manage high shares of renewables (Egging & Tomasgard, 2018). Uncertainty remains over whether battery technology, combined with renewable energy, will develop at sufficient scale and speed to remove gas generally, and LNG's competitive advantage specifically.

Future gas demand forecasts for Europe have been revised downwards owing to decarbonisation targets. Established LNG infrastructure could nonetheless continue to play a key role in ensuring supply security, particularly when it comes to meeting seasonal heating demand, a sector that is challenging to electrify or decarbonise. The marginal cost of gas to provide swing capability is an advantage of LNG in countries where gas is already part of the mix and terminals' utilisation rates can increase usage with less investment. Utilisation rates typically fluctuate on a seasonal basis depending on (1) total market regasification capacity, (2) infrastructure connectivity, (3) liquidity, (4) tradability of the wholesale gas market, (5) competitiveness of regasification tariffs, and (6) attractiveness of the capacity allocation mechanism (IGU, 2021a). Demand can be met in volatile markets when regasification capacity exceeds liquefaction capacity.

Industry participants understood the potential risks that renewables, batteries, and green hydrogen pose for LNG (Al-Kuwari, 2023). Nonetheless, overall seasonality, economics, infrastructure, and scale are areas that LNG can leverage to maintain its role in energy systems over coming decades. Other factors highlighted as potentially limiting the role of renewables with batteries in favour of natural gas and LNG were (1) extraction and disposal of rare minerals, and (2) technical challenges facing the electrification of other sectors in growing economies that need to rapidly reduce emissions (Al-Kuwari, 2023).

Skipping the gas bridge

The potential move away from natural gas is forecast to primarily occur in power generation. Currently, it is taking place mainly in the mature gas markets of North America and Europe. Interviewees discussed the possibility that a more rapid transition could translate to the gas phase being skipped in the traditional evolution from coal to gas to renewables. While coal-to-gas switching may be viable given the overall environmental benefits of gas, the improved performance of coal-fired power plants, the continued promise of "clean coal" technology, and the availability of domestic resources pose a risk to delaying the move away from coal.

Natural gas could face challenges given the pace of development around electrification. Moreover, coal remains affordable and strategically important, particularly in regions where energy demand is forecast to grow the most. LNG has a high cost of production and high investment requirements, whereas domestic coal is generally cheaper and easier to transport. To reach new markets, LNG must continue to compete with coal and renewables in terms of affordability and carbon footprint. "Skipping the gas step" may not be realistic. The logistical challenges and the scale of energy required in power, transportation, and industry are significant. Interviewees also noted that decreasing gas prices close the gap between gas and coal for potential importers (Al-Kuwari, 2023). Historically, LNG prices were approximately three times higher than coal. This premium forces policymakers to make the difficult trade-off between high costs and high emissions. However, there were periods (e.g. 2020 and 2021) where spot cargoes were almost equal to coal, a trend that could significantly increase interest in gas.

The risk of stranded assets

Current and emerging risks could lead to stranded assets. This occurs when "environmentally unsustainable assets suffer from unanticipated or premature write-offs, downward revaluations or are converted to liabilities" (Caldecott et al., 2013, p. 2). Coal and oil appear most at risk of stranding, but natural gas has also received attention in this regard (Teplin et al., 2019). The LNG value chain is a capital-intensive sector requiring financing and sales arrangements that usually span more than a decade. As the pace of decarbonisation increases, participants in the LNG chain must consider the economic viability of LNG projects beyond 2030 to minimise risks of stranding.

Traditionally, LNG projects were underpinned by long-term contracts with a predictable yearly flow of sales through an annual delivery programme. Such programmes factor in seasonality, planned shutdowns, and market dynamics. As LNG imports increased in Europe and North America, the industry moved towards gas-indexed pricing and an increased share of spot volumes as part of the supply mix. This supported LNG's responsiveness to market signals. Several studies have analysed the near- to medium-term impact of additional renewables on commercial models. A modest increase in global CO_2 prices might force gas plants to shift their business model to less frequent, higher peaking, and load balancing operations.

In the United States, nearly 70% of the 70 GW of new gas-fired powerplant capacity announced for 2024 will be uneconomical by 2035 because of improvements in demand-side management and wind, solar, and storage projects. The Rocky Mountain Institute identified (1) carbon pricing, (2) continued decline of clean energy costs, and (3) climate-focused policies towards legacy resources as factors that could increase the risk of stranding gas assets in the United States (Teplin et al., 2019). In the UK, the power market's evolution sheds light on how a heavily gas-dependent energy system introduces significant amounts of low-carbon electricity in a relatively short time. Qadrdan et al. (2015) note that when large shares of nuclear and renewables exist in the generation portfolio, combined-cycle gas plants will increasingly operate as backup supply. This could reduce the gas capacity factor from 46% in 2010 to less than 11% by 2050, significantly impacting the economics of currently operating and future assets. LNG projects have a typical lead time of five years between final investment decision and commercial operation (IGU, 2021a). Thus, uncertainties from the mid-2020s need to be seriously considered.

The long-term risk of falling LNG demand is linked to global energy systems' inability to absorb all recoverable natural gas. The industry has one or two decades of growth, even in a Paris Agreement-aligned world. However, concerns surface beyond the projected peak of gas demand circa 2035–2040: Investment decisions will have to be made at that stage to either extend or replace ongoing projects. In the meantime, companies will be increasingly cautious about investing in new LNG assets, even when there is relative certainty about gas demand for certain markets. Most LNG projects are partly financed on a long-term contract basis, providing a commercial and legal framework to make projects feasible and less risky. It may, however, become more difficult to secure traditional terms associated with established arrangements. That said, if a project is funded, economics will generally be dictated by the marginal cost of supply. The risk of shutting down LNG would then reduce significantly. It is unlikely that a project will not be able to cover its costs and debt once it starts operating (subject to significant events leading to short-term demand destruction). In the context of increasing uncertainty, it was acknowledged that LNG projects with the lowest cost of supply will be the most resilient over the medium to long term.

There is currently no direct link between high emissions and cost. Tangguh in Indonesia, for example, has relatively high emissions but is one of the most competitive projects (Al-Kuwari, 2023). However, as decarbonisation policies develop, carbon price increases could penalise projects. A different dynamic could emerge where the carbon footprint of projects has a direct impact on economics and competitiveness in the long term. This dynamic will increasingly justify investments into decarbonisation measures but might also render some projects uncompetitive.

Interviewees had different perspectives on the risk of stranding depending on the part of the chain they operated in. For shipowners and infrastructure players, the focus centred around the timescale required to recover the value of assets and the ability to pay off debt. For importers, shifting business models in power markets translate into more variability in throughput, affecting their traditional business model and economics. In some cases, buyers will have supply commitments where increasingly uncertain gas demand affects terminal utilisation. Resource holders – a group incorporating sovereign and NOCs – have the longest time-horizon and are most at risk of stranded LNG assets. At the same time, they may be well positioned to capture demand that IOCs traditionally met from their portfolios (should they move away from LNG owing to investment pressure). Resource holders' focus goes beyond the 2040–2050 timescale and centres around the monetisation of resources, which could encourage some players to adopt a more radical approach to decarbonisation.

Decarbonisation policies

Natural gas demand is now forecast to grow at a slower rate than previously thought. It is, however, still too high to be in line with net-zero emission pathways, which push for higher substitution rates and efficiency gains. In the near term, carbon pricing could improve LNG's attractiveness compared with oil and coal. This occurred in the UK, where a carbon price floor of US$25 per tonne spurred a switch from coal. A similar dynamic could take place in Asia with a range of US$25–50. National trading schemes have been implemented in recent years in South Korea and China (Bresciani et al., 2020; Gabbatiss, 2020; GIIGNL, 2020; ICAP, 2022; IEA, 2021a; Liu et al., 2023).

In the long term, global decarbonisation policies will be largely shaped by Europe's approach to LNG because EU countries import more oil and gas than other countries. Changes in the EU's energy policy will have a significant impact on markets (Gustafson, 2020; Stern, 2017). At the end of 2019, the EU launched its Green Deal with the aim of achieving net-zero GHG emissions by 2050 (European Commission, 2019). The EU is likely to have strict carbon measurement criteria for gas imports, including the Carbon Border Adjustment Mechanism for European imports started in 2023. This will initially cover imports of products in carbon-intensive industries. However, it could extend to energy imports in a way that affects LNG exporters in terms of value chain emissions. The European Commission (2019, 2022) has decided that natural gas-fired power plants can qualify for green investments if they meet the terms of the European Green Deal.

Methane is often discussed by policymakers and NGOs and there has been increasing scrutiny of transparency and data quality in the LNG industry. CH_4 is controversial in the climate community because reaching 1.5°C on a 100-GWP basis is inconsistent with the 2050 time-horizon of the Paris Agreement. CH_4 is, therefore, as (if not more) important than CO_2. The EC's Methane Strategy, which took effect in 2024, builds on the standards established under the December 2019 Green Deal (European Commission, 2019). Non-EU natural gas imports are increasing to replace declining indigenous production, contributing between three to eight times the emissions of products produced in the EU (IGU, 2021b). The EU plans legislation on compulsory measurement, reporting, and verification for all energy-related emissions. The creation of the International Methane Emissions Observatory and improved reporting of CH_4 emissions will improve measurement accuracy (European Commission, 2021) and introduce standards covering the full supply chain and the possibility of future legislation on venting and flaring (European Commission, 2024; World Bank, 2020). A default value is proposed for CH_4 emissions from imports that do not comply with the established reporting framework aligned with emissions prices and taxes (Stern, 2020). The European Commission also supports establishing a coalition of major importing countries to coordinate approaches on mitigating CH_4 emissions (European Commission, 2024). The strategy will directly impact the primary LNG exporters into the EU (mainly Qatar, the United States, Algeria, and Nigeria). Importantly, other LNG importing regions are monitoring the EU's efforts and may copy some of the EU policies and strategies (Stern, 2020). In the short and medium term, the EU is focused on producing natural gas-based hydrogen through the application of steam CH_4 reforming and CCS. In the longer term, non-fossil renewable hydrogen (green hydrogen) will be prioritised.

Outside the EU, the Net-Zero Producers Forum was announced between Canada, Norway, Qatar, Saudi Arabia, and the United States in April 2021 with the objective of coordinating zero-emission strategies, including CH_4 abatement (IEA, 2021a). Stern (2020) noted that exporters would likely accept a CH_4 emissions standard for LNG and pipeline export chains, measured and verified from the point of production to the EU border. Politically, CH_4 management is less challenging to address than carbon, which takes on the fossil fuel industry directly. By addressing CH_4 reduction, savings and trade-offs are more aligned between the energy companies and government, allowing for less contentious alignment with the environmental community. Ouki (2023) discusses preparation for a sustainable energy transition in the African context, noting that although GHG emissions in Africa are relatively low compared with the rest of the world, decarbonisation measures imposed by the EU will affect African gas producers. He suggests that addressing associated gas flaring, CH_4 emissions, and the need to abate natural gas supply using CCS or CCUS technology is fundamental for a sustainable energy transition. Ajayi (2024) examined natural gas rents and the qualities of institutions as co-determinants of economic growth among the GECF countries, concluding that the GECF has the potential to rival OPEC in the global energy market, particularly in achieving the UN Sustainable Development Goal number seven. The study suggests that natural gas rent will have long-term positive effects on the growth of the GECF, and that the move towards sustainability in global energy consumption is an economic advantage for the GECF.

Evolving market dynamics: Financing uncertainty

The growth of the LNG industry in recent decades has been underpinned by generally consistent market conditions and a large group of financiers. However, shifting market dynamics and evolving energy systems are creating new uncertainties for LNG participants. The actors involved in LNG financing will have to make decisions based on a dynamic set of variables that affect project economics, including volatile market conditions and global decarbonisation efforts.

From a strategic point of view, long-term LNG financing will be impacted by global sentiments around the role of natural gas, often included in anti-fossil fuel narratives. Historically, LNG was financed by export credit agencies, commercial banks, multilateral agencies, debt capital market agencies, and project owners (Baker, 2020). Referring to multilateral agencies, an interviewee commented that "the stance is turning against natural gas again" despite its apparent importance to developing countries (Al-Kuwari, 2023). Combined with increasing pressure from NGOs, this shift will send signals to commercial banks. Project lenders have tended to follow the evolution of commercial terms agreed upon by sellers and buyers, including a move away from long-term take-or-pay contracts towards more flexible five- to ten-year arrangements. Lenders focus on the overall robustness of the value chain and pricings in target markets. These were originally heavily oil-indexed but there has been a move towards gas-on-gas indexing in certain markets. However, new business models and industry players have introduced complexity to the analysis of financing. The US-based tolling model decoupled upstream resources from the liquefaction process, adding new dimensions of uncertainty. In 2020 – a period of weak LNG demand – US-based projects experienced a disproportionate decline in liquefaction capacity utilisation (from 96.9% in 2019 to 76.5% in 2020). This reflected flexible commercial arrangements in the United States that gave off-takers the option not to lift cargoes (IGU, 2021a). Lockdowns in India, Europe, and Asia limited options for cross-basin optimisation and led to negative netbacks for US-loaded shipments resulting in many cancellations. Most projects in the United States are set to be greenfield projects, consisting of multiple small- to medium-scale LNG trains delivered in a phased way. This approach provides flexibility in managing long-term off-takers requiring traditional LNG financing structures to adapt.

Interviewees noted that overall project return expectations have lowered in recent years. Projects will be "lucky to see [returns] in double figures in LNG across the board" (Al-Kuwari, 2023). The cost of debt for LNG is also increasing, making renewables more attractive because they are associated with less long-term risk. Stern (2017) categorised countries with significant gas demand potential according to high, medium, and low wholesale gas prices to assess whether gas can compete in those markets. Asian countries represented the best opportunities for growth, with a range of US$6–10 MMBTU. Figure 6.8 builds on studies by Stern (2017) and Steuer (2019) to analyse gas import affordability compared with forecasts for LNG project costs in 2025, key for determining whether future projects will attract financing. The focus was on (1) Northwest Europe, (2) Japan, South Korea, and China, and (3) India, Pakistan, and Bangladesh. Qatar, the lowest-cost supplier, is resilient across all markets, including low-income markets below US$6/MMBtu. North American and Mozambican suppliers have project costs ranging from near US$6/MMBtu to well below the US$6/MMBtu threshold for South Asia and Northwest Europe. Only Nigeria could realistically deliver to South Asia (Indian, Pakistani, and Bangladeshi markets). The margin is, however, very slim. Nigeria – with Africa's largest natural gas reserves and population, and a comparatively large and diversified economy – should play a key role in the continent's future gas supply (Energy Capital & Power, 2024). The country has the potential to dominate LNG development in West Africa but faces persistent challenges, including gas production and infrastructure problems in providing adequate supply to its domestic and export markets (Ouki, 2023). North American projects struggle to meet thresholds for Northwest Europe, and Pakistan, India and Bangladesh because of high overall costs.

Some energy majors categorise LNG as resilient to climate-centred scenarios, and have, therefore, increased strategic investments along the LNG chain (IEA, 2020). Natural gas, particularly LNG, is viewed as a portfolio tool to reduce overall

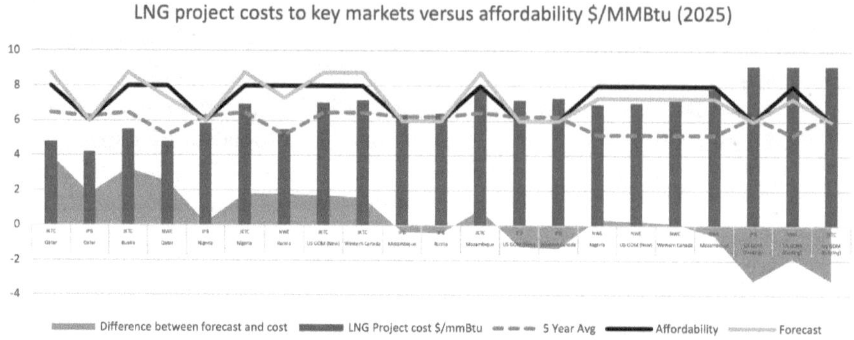

Figure 6.8 LNG project costs (JKTC: Japan, South Korea, Taiwan, and China; NWE: UK, The Netherlands, and Belgium; IPB: India, Pakistan, and Bangladesh) (Steuer, 2019; Stern, 2017)

carbon intensity. It constitutes a key step to reducing scope 1, 2, and 3 emissions. Scope 1 and 2 are direct emissions associated with the supply of heat, electricity, and cooling, while Scope 3 are value chain emissions not directly related to the purchase of energy. In the supplier's case, they are indirect or "collateral emissions" (BEIS, 2020; Fong et al., 2021). Shell, Total Energies, and other international energy companies forecast an increase in LNG exposure as part of their net-zero objectives (Royal Dutch Shell, 2021; Total Energies, 2022) leading to increased investment in the sector in the short to medium term. However, companies that have adopted aggressive measures to reduce their carbon intensity have less scope for new LNG projects in the long term. Going forward, the focus will be on reducing the emissions of existing projects and/or rebalancing portfolios. The industry needs to develop strategies beyond the immediate phase of lowering overall portfolio carbon intensity. Companies must assess how they can continue to invest without wasting money. One executive asked:

> What will happen? Will LNG still be a part of the mix? Or can we envisage a future where we've transitioned away from fossil fuel solutions? . . . that is a dilemma, and the question that all suppliers, producers of LNG are asking themselves right now.
>
> (Al-Kuwari, 2023)

Government and NOC approaches to the viability of LNG projects focus more on monetising upstream reserves than on maximising short- to medium-term returns. NOCs and SOEs vary in terms of their ability to finance their own projects. In most cases, they will require funding support from companies or lending institutions. According to the IEA (2020), SOEs represent less than 10% of all energy investments in advanced economies compared with over 50% in developing economies.

Market conditions

Several factors have contributed to an increase in boom-and-bust cycles in LNG, further contributing to uncertainty. Few major project announcements have been made in the LNG industry since 2016 and the long lead times of LNG project developments accentuate the disparity between near-term market conditions and new project announcements (IEA, 2020).

The shock to the global economy caused by COVID-19 led to short-term over-capacity, in turn contributing to decreased margins and the cancelling of planned investment. Nonetheless, the industry anticipates a shortfall of LNG supply by the mid-2020s. Equity investors are taking a proportionate share of LNG volume to market instead of relying on the traditional long-term agreement-based structure to lock in demand. Longer cycles in LNG investment will lead to over-supply, making it difficult for the industry to respond to market dynamics and adding additional risks for exporters and investors.

Regarding pricing structure, most LNG contracts are linked to Brent crude oil prices. Older contracts are near the average slope of 13% to the dated Brent. Newer contracts signed in 2020 are near an 11% slope, correlated with the fall in LNG spot prices in the first half of 2020 (IGU, 2021a). In Asia, the lack of a single connected gas market will prove to be a challenge for developing an alternative. Nonetheless, the Japan Korea Marker (JKM) is developing as a principal regional index despite not being a traditional gas hub. Trade has increased in recent years, and lenders will have to accept LNG as a viable alternative for a portion of traded volumes. LNG pricing must remain attractive to meet financing thresholds but must also be able to compete with alternatives. An interviewee noted that the LNG industry must move past pricing LNG at a premium (Al-Kuwari, 2023). LNG must position itself to price out more polluting fuels – including coal and kerosene – while reducing its carbon emissions. Lenders and the industry may increasingly have to consider radical approaches, including fixed pricing to deal with increased complexity and resulting uncertainty. Relatively high LNG prices are especially discouraging in India and Southeast Asia, where US$10/MMBtu is clearly not competitive (Al-Kuwari, 2023).

Qatar and Russia, both GECF countries, are making strategic attempts to capture market share, with the potential to price out higher cost supplies (IEA, 2020). Qatar, one of the largest LNG producers, has a unique strategic impact in the industry. It is the most competitive source of LNG supply, and is economically supported by high condensate yield when producing natural gas. This directly improves project economics and offers flexibility for marketing strategies. Qatar Petroleum took a final investment decision for North Field East (NFE), raising the country's production capacity to 110 MTPA, overtaking Australia and the United States as the country with the most liquefaction capacity (Global Energy Association, 2023). Qatar has also announced plans for an additional 16 MTPA from the North Field South (NFS) project, effectively increasing Qatar's LNG production to 126 MTPA by 2027 (GECF, 2024; IGU, 2021a). The US$30 billion Qatar expansion includes emission reduction initiatives such as utilising power supply from the national grid, sourcing a portion of the power required from an 800-MW solar PV power plant, and an integrated carbon capture project aimed at capturing 7MTPA of CO_2 by 2030 (GECF, 2023; IGU, 2021b). The Qatar LNG expansion and proactive efforts to decarbonise LNG's value chain may be "game changers" that alter the LNG financing landscape. Qatar Petroleum had a moratorium in place on gas expansion until the latest announcement. Interviewees interpreted the announcement as a recognition by Qatar Petroleum that the role of gas in the energy transition is shifting. Qatar has approximately 100 years of gas supply remaining at current output levels. The challenge now is to maximise the value of that resource for the next 30 years and invest accordingly.

Societal pressure

The industry has missed opportunities to make a better environmental case for LNG to the public, investment community, and climate activists. More proactive public relations could have dissociated LNG and natural gas from other fossil fuels. A "gas is good" message was not consistently promoted. Instead, there has been a tendency to rely on a "gas is better than coal" argument. The industry did not make enough effort to position LNG as a pragmatic solution as part of the pathway towards a low-carbon economy. This was especially the case 10–15 years ago when there was a realistic window to engage policymakers and be more active in climate discussions.

The LNG industry could also have improved its image by increasing transparency around its emission profile, positioning itself as openly and proactively discussing decarbonisation in ways the LPG, oil, and coal industries are not. The LNG industry should be portrayed as actively monitoring and reducing its carbon footprint, and should rapidly coordinate its decarbonisation efforts. LNG should be included in "fair and just" transition narratives as providing short- to medium-term support for developing countries with significant electricity demands (Al-Kuwari, 2023). Developing countries have limited options, and LNG provides the fastest route to development. Grid and transmission issues, intermittency, and the need to electrify on a large scale might lead to coal maintaining or even growing its role. LNG is a lower-carbon alternative. The following section focuses on the LNG industry's options moving forwards.

Technical analysis of the LNG industry

Transformation: Hydrogen

The LNG industry has the option of transforming itself into a hydrogen industry. In a series of studies exploring decarbonisation and the natural gas value chain, Stern (2017, 2019a, 2019b, 2020) highlighted how the LNG industry should deal with rapidly decarbonising markets. The industry should address the issue of CH_4 emissions, increase transparency, and even contemplate measures like shipping blue hydrogen instead of LNG. The risk of not doing so is that carbon – and possibly

CH_4 – emissions will cause natural gas to be "progressively unburnable" (Stern, 2017, p. 1). According to the IEA, "clean hydrogen is currently enjoying unprecedented political and business momentum" (2019, p. 13). This differs from previous periods owing to hydrogen's flexibility in meeting complex energy-transition challenges and wide support from government and industry (IEA, 2019). In the 1970s, oil price shocks coupled with increased emphasis on curbing air pollution heightened interest in hydrogen. Growing environmental awareness further highlighted its potential benefits in the 1990s. However, oil prices remained too low to encourage the relevant projects. In the early 2000s, a continued focus on climate change reintroduced hydrogen as a potential solution. However, hydrogen development was pushed back by a lack of alignment with policies, increased oil and gas production, and improvements in battery technology (IEA, 2019). Thus, the high expectations for hydrogen and fuel cells that characterised the 2000s did not materialise. In fact, general government spending on hydrogen declined by approximately 35% between 2008 and 2018 (Van de Graaf et al., 2020).

The driver for hydrogen development in the twentieth century was its potential economic competitiveness with hydrocarbons. Now, it is seen as a key tool for combatting climate change (IEA, 2023a; Mitrova et al., 2019), although this may alter in the future with improvements in technology, electrolysers, declining renewable energy costs, increase of commercial products, global alignment to combat climate change, and recognition that electrification through renewable energy cannot address all sectors of the economy (Staffell et al., 2018; Van de Graaf et al., 2020). With declining renewable electricity costs, interest in electrolytic hydrogen has increased in recent years (IEA, 2019, 2023a, 2023c).

The impact of hydrogen trade on the LNG industry

Al-Kuwari and Schönfisch (2022) studied the impact of increased low-carbon hydrogen production and consumption on LNG flows, and explored synergies between hydrogen and LNG value chains in the context of decarbonisation. They developed a novel, integrated natural gas and hydrogen market model for several scenarios, all in the context of aggressive global decarbonisation, to analyse potential global natural gas and low-carbon hydrogen development pathways until 2050. In technology agnostic scenarios, LNG demand remains resilient until 2050, even as overall natural gas demand decreases. In scenarios where green hydrogen becomes dominant, LNG demand begins to decline around 2040. These projections are supported by the cost advantages of natural gas-based (NG) hydrogen over renewable energy sourced (RES) hydrogen.

Production pathways and natural gas demand could be significantly impacted by differences in gas prices and/or changes in renewable energy costs and electrolysers. Al-Kuwari and Schönfisch also found that it is more economical to produce hydrogen locally. For gas-based low-carbon hydrogen, transporting gas is more efficient than hydrogen. In contrast, for RES-based hydrogen, transport would take place only via regional pipelines in certain situations. LNG producers would be the primary beneficiaries of an increase of the natural gas-based hydrogen demand. It is likely that most low-carbon hydrogen will be consumed in the advanced economies of the Asia Pacific region owing to their reliance on LNG imports (Al-Kuwari & Schönfisch, 2022).

Hydrogen and the LNG value chain

The hydrogen value chain has the potential to evolve in coming decades. Table 6.3 presents an analysis of the wider set of strategic issues that impact LNG exporters.

Studies have explored the technical and operational links between hydrogen and LNG, focusing on production methods, shipping, and storage (Abe et al., 1998; Al Ghafri et al., 2023; Bang et al., 2011; Hanley et al., 2015; Lloyd's Register & UMAS, 2019). Even though hydrogen requires significantly colder temperatures to become a cryogenic liquid (–253°C compared with –161°C for LNG), there are potential synergies to explore, including using LNG for pre-cooling in hydrogen liquefaction to improve efficiencies. Liquefying hydrogen increases overall production costs, but can lower transportation costs, although this process would consume 25–35% of the hydrogen compared to 10% for natural gas (Australian Renewable Energy Agency, 2018; Bruce et al., 2018; IEA, 2019). Industrial clusters, the site of many LNG import terminals, could also provide a unique opportunity for LNG to be imported and for hydrogen to be produced at the same location (IEA, 2019). Coastal industrial clusters with large dependable industrial customers represent primary near-term opportunities where existing gas infrastructure can support the scale-up of low-carbon hydrogen production and consumption. Blending natural gas with low-carbon hydrogen in pipelines supports the introduction of hydrogen as an energy carrier, and reduces emissions attributed to natural gas. However, when it comes to shipping, a University Maritime Advisory Services study found that LNG bunkering infrastructure is unlikely to be suitable for liquified hydrogen (LH_2). Retrofitting could be as expensive as building new infrastructure because, even though both LH_2 and LNG require cryogenic treatment, they require different cooling equipment and insulation (Lloyd's Register & UMAS, 2019).

Several possible importers of hydrogen are already LNG importers with established links to LNG exporters (e.g. Japan and Korea), and therefore, possess transferable commercial and value chain management knowledge. The hydrogen industry could develop similarly to the LNG industry through establishing a hydrogen market based on bilateral long-term contracts supported by government-to-government agreements, and take-or-pay commitments to underpin investment into hydrogen production, storage, and transport assets (Bruce et al., 2018; Van de Graaf et al., 2020). Initial supply agreements would also benefit from negotiating favourable trade tariffs, supported by joint ventures that leverage current commercial relationships and share risk through a vertically integrated approach. For example, Australia is an important LNG exporter that looks to leverage its LNG position and expertise to support its hydrogen export plans. It has developed a national hydrogen strategy, currently under review, initially focused on electrolysis, and then on the large-scale production of hydrogen using brown coal (Bruce et al., 2018; Van de Graaf et al., 2020). The roadmap developed by Australia's national

Table 6.3 Strategic issues impacting LNG exporters (Al-Kuwari & Schönfisch, 2022)

Strategic choices			*Impact on LNG exporters*
Value chain options	*Natural gas*	*Renewable energy*	
Production technology	Natural gas reforming or pyrolysis	Electrolysis	• LNG exporters can play a direct role by delivering the natural gas feedstock needed • Minimal opportunities for LNG exporters in green pathways
Production location	Domestic gas to be converted to hydrogen	Hydrogen produced via renewable energy and electrolysers domestically	• LNG exporters could convert natural gas into hydrogen at LNG-receiving terminals, subject to CCUS access or CH_4 pyrolysis technology
	LNG to be imported and converted to hydrogen near LNG-importing terminal	RES-based hydrogen imported via liquified hydrogen, LOHC, ammonia, or derivatives	• For import-based strategies, there is a limited role for LNG. Exporters can leverage know-how and commercial links • For pure non-import strategies, there is no role for LNG exporters
Hydrogen transport (pure hydrogen versus derivatives)	Natural gas can be converted to nearly pure hydrogen with pyrolysis or CH_4 reforming and CCUS	Pure hydrogen produced via low-carbon renewable energy.	• Using current LNG infrastructure and import terminal allows LNG to be shipped and converted into hydrogen at importing countries
	Natural gas could be converted into synthetic methanol, CH_4, Fischer–Tropsch liquid hydrocarbons, ammonia, and so on	RES-based hydrogen could be converted into synthetic methanol, CH_4, Fischer–Tropsch liquid, hydrocarbons, ammonia, and so on	• Pure hydrogen and other derivatives have different shipping and receiving requirements. This limits the scope of LNG exporters

(*Continued*)

Table 6.3 (Continued)

Strategic choices			*Impact on LNG exporters*
Value chain options	*Natural gas*	*Renewable energy*	
Hydrogen end-use (targeted versus economy-wide)	Natural gas-based hydrogen can target applications, such as chemical feedstock, oil refining, steel production, and ammonia production. It can initially replace current hydrogen demand	RES-based hydrogen to target applications such as chemical feedstock, oil refining, steel production, and ammonia production. It can initially replace current hydrogen demand	• LNG terminals located near industrial clusters present a unique opportunity to take advantage of technical synergies • For wide-scale hydrogen adoption, importers could leverage existing gas infrastructure by blending and similar measures in the medium term. LNG exporters will depend on favourable government policy and on the ability to manage CO_2 (carbon shipping, pyrolysis) • Countries may take political decisions to utilise only RES-based hydrogen. This would significantly limit LNG exporters' role
	Natural gas-based hydrogen's role in large-scale hydrogen adoption will depend on technical characteristics, economics, and government policy	RES-based hydrogen's role in large-scale hydrogen adoption will depend on available renewable resources, economics, and government policy	
Hydrogen role (exports versus domestic use)	Natural gas-based hydrogen supports exports to sectors and countries with increasing hydrogen demand	RES-based hydrogen allows for countries with rich renewable energy resources to export a new commodity (subject to geographical location and technological improvements)	• LNG exporters could continue to export to end-users with aggressive decarbonisation targets • LNG exporters who are resource owners may consider shifting to exporting hydrogen, reforming natural gas, and managing carbon domestically. This would alter the entire value chain
	Natural gas-based hydrogen could decarbonise domestic hydrogen and other derivatives	RES-based hydrogen could decarbonise domestic hydrogen and other derivatives	

LOHC – liquid organic hydrogen carrier; RES – renewable energy source.

science research agency (CSIRO – the Commonwealth Scientific and Industrial Research Organisation) states that lessons can be learned from the LNG industry, including mimicking the origins of LNG market development by leaning on government-to-government agreements to support long-term off-take contracts to secure sufficient and stable financing.

Natural gas-based low-carbon hydrogen production can play a major role in supporting LNG exports. However, model-based scenario analysis suggests that directly exporting pure hydrogen is not a viable option for most LNG producers and exporters. While there may be potential synergies and transferable knowledge, there are limited incentives for LNG exporters to become hydrogen exporters (thereby fully decarbonising the value chain). Actors could respond and adapt to hydrogen development differently. For countries where gas revenue represents a significant source of income, increasing hydrogen demand presents a long-term risk for their natural gas. Increased investment in hydrogen and CCUS technology, and expanding LNG terminals to explore synergies with industrial clusters, might ensure that future strategies are in line with decarbonisation policies. A long-term low-carbon strategy also provides an opportunity for LNG resource owners to support their host governments' energy-transition-related diversification plans, increasing their ability to monetise their natural gas reserves while maintaining access to high-value markets.

LNG exporters would benefit from adopting natural gas-based low-carbon hydrogen in import markets. The interviewed LNG industry participants acknowledged the growing importance of low-carbon hydrogen and identified hydrogen as both an opportunity and a threat (Al-Kuwari, 2023). The LNG industry is well positioned in terms of skills and resources to play an important role in developing low-carbon hydrogen.

Green hydrogen

Low-carbon hydrogen is recognised as a technically viable tool for addressing long-term decarbonisation targets. Green hydrogen – hydrogen fuel produced using renewable energy – may be particularly threating to the LNG industry because it can provide direct competition to both natural gas-derived hydrogen and current CH_4-based solutions. Moreover, low-carbon hydrogen is increasingly part of decarbonisation narratives related to LNG, forming part of more radical solutions for gas resource owners to decarbonise their entire chain. LNG industry interviewees were, however, sceptical of low-carbon hydrogen's ability to play a significant role in energy systems in the medium term (Al-Kuwari, 2023).

Current hydrogen demand of 69 million tonnes – derived almost entirely from fossil fuels – is likely to be the initial target of low-carbon hydrogen supply and is most likely to be met by natural gas-based hydrogen, namely blue hydrogen. Green hydrogen was characterised as "still at a very early stage, with very low adaptation rates, and a slower development curve than people might think" (Al-Kuwari, 2023). Interviewees mentioned technical challenges related to transportation, infrastructure, and the cost of supplying low-carbon hydrogen. LNG's value chain is relatively straightforward compared with hydrogen's fragmented chain. Sustained global support (similar to that received by renewables in Europe) and the support of other technologies (like CCS) are needed to scale up low-carbon hydrogen. In the IEA's Sustainable Development Scenario (SDS), global capacity for electrolysers needs to expand dramatically to reach net-zero emissions.

Low-carbon hydrogen demand will differ by industry and region, partly owing to low-cost renewables. Even if potential renewable resources exist in developing countries, meeting domestic power demand will be the priority in upcoming decades. When referring to aggressive low-carbon hydrogen forecasts in Europe, an interviewee pointed to the slow growth and uptake of electric charging stations in the UK as an example of reality falling short of expectations. Hydrogen as a carrier is even more complex to manage than electricity, and presents additional safety concerns. Low-carbon hydrogen was seen as both an opportunity and a threat given LNG's potential to play a role in blue hydrogen pathways (Al-Kuwari & Schönfisch, 2022). However, green hydrogen was seen as a serious threat beyond 2040, especially if domestically produced or if long-distance hydrogen shipping becomes economically viable. Advances in electrolysis technology, cost improvements, and potential CCS challenges could reduce the window for blue hydrogen to play a role.

International oil companies and fossil fuel producers are exploring ways to repurpose infrastructure and assets (Al-Kuwari, 2023; Al-Kuwari & Schönfisch, 2022). Pyrolysis was identified as a potential "game changer" if the technology can be commercialised at scale by 2030. The current level of readiness is, however, not very high. Technologies that are most suited for large-scale use are not the most advanced. Thus, the LNG industry could be significantly impacted if green hydrogen pathways gain momentum, but only in the long term. In the meantime, it is "definitely blue [hydrogen] all the way through" (Al-Kuwari, 2023), and LNG could be the primary feedstock for this development.

Optimisation: Management and options

Determining the true emissions profile of the LNG chain is an important step that the industry needs to take to support gas switching and decarbonisation narratives. Along with the lack of transparency regarding emission profiles, assessing the overall net impact of LNG imports in energy systems in transition has become increasingly complex. Several mitigation measures exist for each value chain component, with varying degrees of impact and maturity. CCS can play a role in capturing reservoir CO_2 at the liquefaction and power plant level. Venting and CH_4 leak reductions are key in the upstream and regasification phases. Electrification can be applied in the upstream at the compressor stations, replacing gas turbines at the liquefaction and power plant level. Reduction of CH_4 slips and optimising shipping fuels are key in the transportation phase (IEA, 2021a). When it comes to addressing emissions, an increasing number of companies in the LNG chain are developing an integrated approach. Stakeholder interviews revealed key themes regarding the impact of decarbonisation methods (Al-Kuwari, 2023), discussed in the following sections.

Value chain emissions responsibility

There are three broad categories of GHG emissions in the LNG value chain – upstream, liquefaction, and point of use. Players range from full chain owners (including NOCs and large integrated energy companies) to those that operate in one or a few distinct parts of the chain. Other players include upstream gas producers, natural gas pipeline operators, liquefaction companies, regasification terminal owners, LNG importers, pipeline gas importers, shipowners, and ship charterers. Traders, consultants, service providers, and a range of other participants also support the LNG chain ecosystem. A company's role in the LNG "ecosystem" will impact its approach to managing emissions.

Responsibility for oil and gas emissions ties into a larger debate around scope 1, 2, and 3 emissions. Interviewees noted that, while regulations and markets will determine the industry's overall responsibility, companies will likely only be held accountable for emissions they can actively measure and control. The upstream will likely be governed by host country regulations, while the downstream will depend on commercial and environmental factors. Commercial players will, however, have distinct responsibilities within the chain they participate in. An example of this dynamic is a growing US LNG commercial model centred around tolling. Here, LNG players cannot manage the emissions of the upstream, but can select their supplies from lower carbon intensity fields. While they are not responsible for transporting natural gas to their LNG facilities, they can introduce incentives by selecting suppliers with strong environment track records and CH_4 management potential. Once the natural gas is received, emissions can be lowered by CCS to manage CO_2 and low-carbon electrification of liquefaction facilities. These are options under the liquefier's control.

"No regrets" measures

Several categories of emission measures exist, each with economic and strategic implications. Improving operational efficiency, liquefaction operations, and electrification were widely supported and considered to be "no regrets" measures. An industry-led and coordinated effort to establish standards for emission measurements and value chain methodologies will support decarbonisation decisions. Information gathering and certification could also be used as tools to demonstrate that the industry is operating according to agreed standards. This would assure governments, buyers, and the public that LNG is produced, shipped, and delivered responsibly. CH_4 leakage management is key for operational efficiency, and efforts are needed to manage this at every stage, for example, by using satellites and leak detection and repair (LDAR), capturing vented gas, upgrading pneumatic devices, and replacing gas-powered pumps with electric pumps across the chain. Where possible, companies should explore using lower-carbon electricity rather than fossil fuels, including electrifying the power to the plant, either via the power grid or on-site renewables. The aim is to reduce the need for feed gas as much as possible. Electrification also presents an opportunity to run existing facilities rather than using pneumatic pressure, diesel, or gas. Companies have taken steps to reduce their expected emissions, particularly from valves and gas-powered pneumatic devices. Common options include installing new emissions control devices, like vapor recovery units (VRUs), and replacing pneumatic devices with air instrument systems. VRUs capture gas accumulated in oil storage tanks that would otherwise be vented into the atmosphere (IGU, 2021a, IEA, 2023c).

Abatement investment decisions

Uncertainty around how to address emissions is prevalent among LNG players, partly because of evolving regulations and shifting decarbonisation targets and because some impactful measures require significant capital investment and complex, long-term strategic considerations. Each project will have different technical and geographic characteristics. Participants need to make economic and technical trade-offs to invest in solutions that reduce emissions most effectively. Project life and well characteristics are major factors in potential investment decisions, where new projects will be increasingly evaluated through the energy-transition lens. CCS is key for reducing liquefaction and combustion-related CO_2 emissions that comprise up to 25% of upstream emissions. CCS continues to be a high-cost solution with unclear market economics. As such, several projects – including Snøhvit and Qatar – have announced that they will integrate CCS at their facilities (Gilbert & Sovacool, 2017; Wood Mackenzie, 2020b).

Electrification can also be applied on a large scale. The Wood Mackenzie energy research and consultancy group (2020a) reports up to 8% of emissions could be reduced by using renewable energy with battery storage to power LNG plants in Asia Pacific. One interviewee pointed to Equinor's efforts to use offshore wind-powered electricity to run facilities. While this is a costly option, project sponsors must compare it to more complex and expensive carbon capture and injection options. Other benefits of electrification would be increasing the amount of feed gas that can be commercialised, and capitalising on potential benefits when carbon taxes are factored in (Wood Mackenzie, 2020a). Importers and traders may increasingly rank suppliers according to assessments of exporter Environmental, Social and Governance (ESG) policies, carbon intensity of different sources, and adopted decarbonisation measures. Policies like the EU's proposed Carbon Border Adjustment could also impact the economics of importers and exporters. Furthermore, there is an opportunity for LNG to support GHG reductions for buyers, commanding a premium if monitoring, reporting, and verifying requirements are implemented. Figure 6.9 depicts the flow of LNG to power plants and associated emissions.

In conjunction with technical measures, LNG actors can take commercial and strategic steps to deal with the impact of decarbonisation. Commercial measures could include full value chain emission measurement and certification or leveraging industrial clusters by incorporating LNG projects with potential hydrogen uses (e.g. CH_4 reforming or pyrolysis). Strategic measures could include radical steps like decarbonising the entire LNG value chain with a focus on offsetting emissions

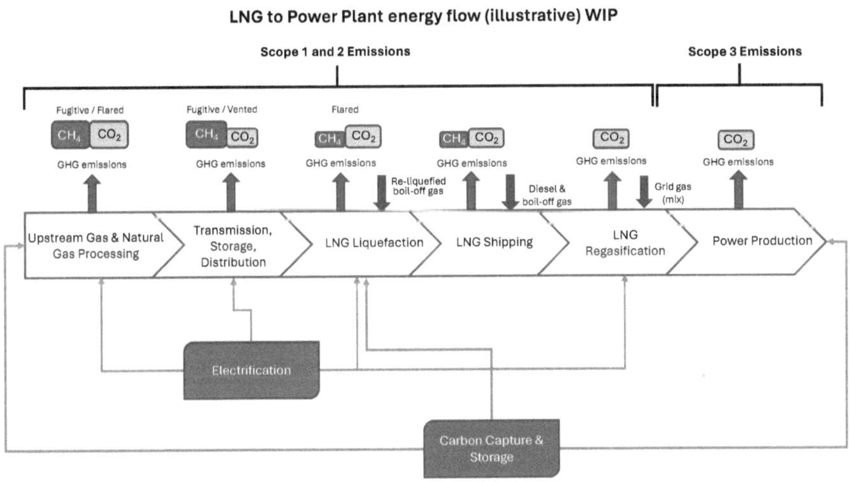

Figure 6.9 LNG to power plant flow and emissions (adapted from Roman-White et al. (2019)

via carbon-neutral cargoes and hydrogen – an emerging trend in the industry that represents a potentially more structural competitor and even a possible solution to decarbonisation.

Carbon-neutral cargoes and offsets

The International Group of Liquified Gas Importers (GIIGNL) defines carbon-neutral LNG cargoes as shipments where the GHG emissions have been offset through the purchase of credits from carbon removal projects, for instance, projects linked to afforestation, reforestation, or renewable projects (GIIGNL, 2020, p. 5; Wood, 2024). Carbon or GHG offset is an emerging decarbonisation tool, widely supported to promote the industry's efforts at decarbonising the LNG chain. By the end of 2020, up to 20 out of 5,000 sold cargoes were carbon neutral (Blanton & Mosis, 2021; Bose et al., 2021). The vast majority were imported by Asian companies, where a coalition of fourteen Japanese end-users formed the Japanese Carbon Neutral Buyers LNG Alliance (Timera Energy, 2021). Companies involved in these transactions are willing to invest in carbon emissions-related trading activities (Blanton & Mosis, 2021; Stern, 2020; Wood Mackenzie, 2020b). The relevance of carbon-neutral cargoes highlights many of the challenges related to decarbonising the LNG chain. These include the availability of emissions data, consistency in methodology, and impact on economics. Consideration also needs to be given to the parts of the value chain to be measured and included in offset calculations given the lack of industry consensus on the matter. Calculating the product life-cycle emissions of each cargo is challenging. Characteristics of the upstream source, LNG technology, transportation methods, and downstream end-users all must be considered. The lack of standardisation has prompted several industry participants (including Pavilion Energy and Cheniere) to develop methods and reporting standards across their supply chains (Timera Energy, 2021).

A benefit of LNG is that each part of the chain can be measured and accounted for separately, providing options for offsetting. However, methods for calculation can differ significantly depending on which part of the chain the company participates in. Assessments are more challenging for firms who procure gas from multiple suppliers, locations and pipelines. This contrasts with resource owners who have an integrated view of the emission profile and are, thus, better positioned to establish systems informed by accurate data. Producers and exporters have direct control of the emissions up to the point of delivery. The importer is in control of measures related to the use of LNG. Buyers and sellers negotiate and make strategic decisions to determine which parts of the supply chain should be part of the calculation (McHich & Britnell, 2021). Carbon offset of an LNG cargo can be applied to reflect emissions of various parts of the chain, either the full life cycle from upstream production to end-user combustion or specific parts of it (Blanton & Mosis, 2021; GIIGNL, 2020, 2021; IEA, 2021a). Other examples highlighting the wide range of contrasting options include scope 1 to scope 3, CO_2 well-to-wheel, CO_2 end-use combustion, CO_2e well-to-wheel, and GHG well-to-wheel (IEA, 2021a). In terms of offsetting, the type of mechanism selected depends on several factors, including project location and the relevant sponsors' objectives. Offset pricing mechanisms include certified emissions reductions verified under the Kyoto protocol, the Verified Carbon Standard (VCS) backed by forestation, renewables, and emissions reductions projects, and participants' own portfolios of nature-based solutions and projects.

At this early stage, uncertainty and variability are high owing to the confidentiality of commercial transactions. Along with agreement on which part of the chain should be offset and the mechanism of offsetting, there also needs to be agreement on which party pays the "price premium". Some buyers may be prepared to accept a premium to manage public scrutiny. However, in the long term, this dynamic will reflect the size of the incremental cost. Offset cost is the other variable – typically wide-ranging and uncertain in the absence of industry standards (Blanton & Mosis, 2021; IEA, 2021a; Wood Mackenzie, 2020b). The overall amount of the carbon-neutral cargo premium is the total emissions of the cargo, the component of the value chain it includes, and the offset cost. Carbon-neutral cargoes are based on the premise that all emissions not reducible or removable along the LNG chain will be offset. The premium is a proxy for how carbon-efficient LNG is. In the long term, lower intensity projects are favoured, and owners incentivised across the chain to invest in decarbonisation efforts. In practice, upstream shipping and regasification actors can minimise the emissions of their components. End-users will have to manage most of the emissions downstream. This leads to the bigger question of who is responsible for the overall decarbonisation of LNG. Going forward, environmentalists will likely stress that carbon-neutral cargoes do not contribute to actual emission reductions (Greenpeace, 2023). They will likely push for a wider definition, where the full value chain is accounted for so that all emissions (whether direct or indirect) are included.

While there is an expectation of increased carbon-neutral cargoes to scale up, achieving significant impact on emission reductions is unlikely. An average-sized cargo emits approximately 270,000 tonnes of CO_2 equivalent over its life cycle (Blanton & Mosis, 2021; Wood Mackenzie, 2020b). Scaling up will not be achievable economically if, as expected, offset prices increase with increased demand. Offsetting is currently popular and attractive because it is sufficiently cheap for suppliers to pass the price on to buyers, or for traders and sellers to create an option and manage the process. While buyers are satisfied with the "green" LNG cargo label, demand will become uncertain as prices increase.

Some carbon-neutral cargo transactions based on public information and industry analysis for the period 2019 to mid-2021 have been publicly disclosed (Blanton & Mosis, 2021; Bose et al., 2021; McHich & Britnell, 2021). Carbon-offset costs range from US\$0.32 to US\$51 per tonne of CO_2. This translates to a range of several thousand to several million dollars per cargo. Neutralising the carbon of LNG cargoes can significantly increase overall costs depending on transportation costs and market prices (IEA, 2021a; Procaccini & Markowitz, 2020).

In the full emissions scenario, the full emissions of a cargo are offset. In the well-to-tank scenario, all emissions up to the point of delivery are included. Key assumptions include (1) a cargo of 145,000 m^3, (2) 62,250 tonnes of LNG delivered, (3) 237,510 tonnes of CO_2e emissions for full emissions, (4) 57,420 tonnes of CO_2e partial emissions, and (5) average carbon-offset prices ranging from US\$3 to US\$39 in developed and emerging markets for 2025 (Bose et al., 2021). The emissions calculations are based on (1) 3.42 of CO_2e generated for each kg (for the full value chain, including combustion), and (2) 0.88 of CO_2 equivalent generated for each kilogram for the extraction, refining, and transporting LNG. These estimates are based on the UK's BEIS (2020) CO_2e emissions allocation and conversion factors. These emissions include all those associated with extraction, refining, and transporting LNG imported into the UK. These figures are not chain-specific and provide broadly similar ranges to the studies referred to above, where approximately one quarter of emissions are well-to-tank related (excluding combustion).

For levels seen in recent compressed natural gas deals, the offset costs can add up to US\$0.50–0.75/MMBtu per cargo. This can total up to US\$2.5 million per cargo at a range of 5–10% of the total cost of a cargo (Blanton & Mosis, 2021; Bose et al., 2021). As offset prices increase, cargo emission neutrality becomes more challenging and full emission neutrality quickly becomes unsustainable. A similar analysis was undertaken by Bose et al. (2021), who calculated the carbon-offset credits as premium. Here, the offset costs as a percentage of the total cargo value are calculated using prices from Steuer's 2025 (Steuer, 2019) forecasts between US\$6 and US\$9 and the offset ranges highlighted by Bose et al. (2021) of up to US\$50, demonstrating the impact of levels of up to US\$100.

Carbon-neutral cargoes are at an early stage and could become more widespread as their adoption gains momentum (e.g. their appeal could extend beyond Asia to Europe). However, European regulators will likely only consider them a transitory tool and will continue to focus on decarbonising the value chain. Opportunities also exist for innovation around carbon-neutral cargoes coming out of the

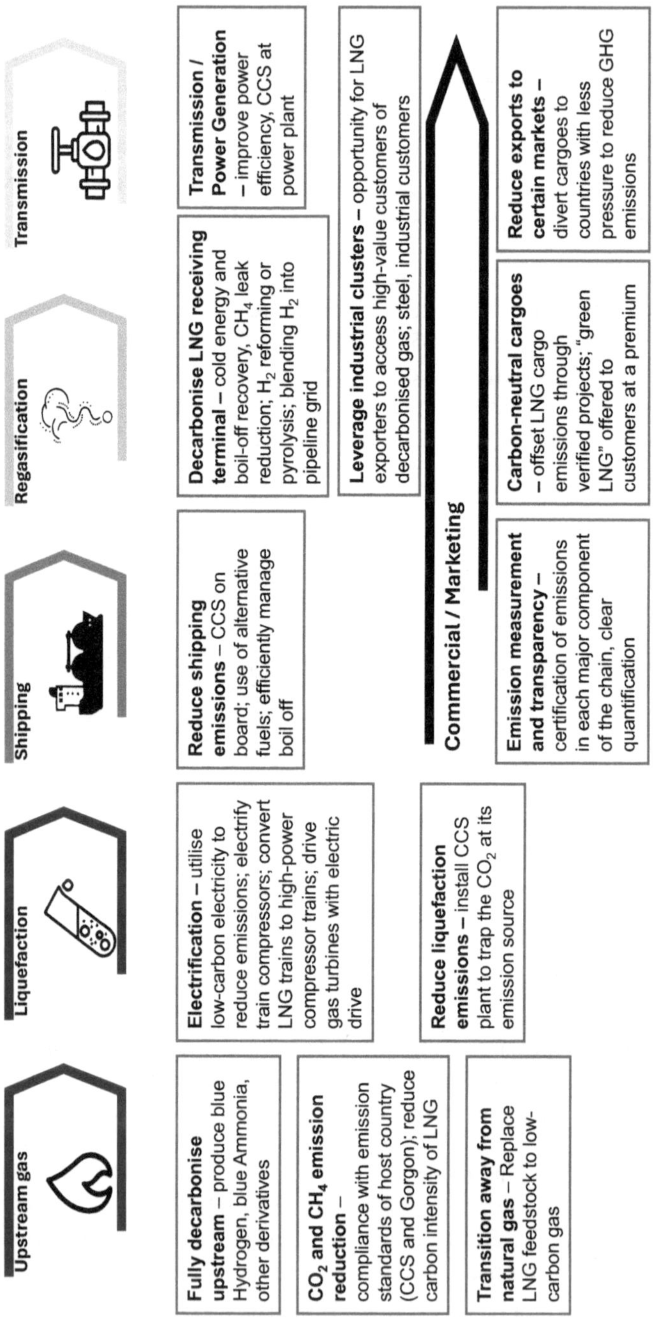

Figure 6.10 Options for decarbonisation

US marketplace and blockchain technology that supports commodity trading and validation for customers. In the long term, carbon-neutral cargoes might become a "drop in the ocean" or even "window dressing" (Al-Kuwari, 2023). They are currently part of the decarbonisation toolbox and helpful for differentiating LNG from competing fuels in the short term. However, the industry should ultimately focus on reducing its carbon footprint along the chain. If the system is not more transparent, there is a risk that the above-mentioned efforts will backfire and be labelled "greenwashing" by NGOs.

Based on this analysis, carbon neutrality may not be a viable long-term solution for the industry, especially when the full value chain is considered. As such, the key factor going forwards for the LNG industry will be sustainable reductions of emissions through operational and technical measures. In principle, hydrogen combined with CCS presents an opportunity for the LNG value chain to fully decarbonise. Figure 6.10 provides a visual summary of options for LNG decarbonisation.

The impact of CH_4 on the natural gas value chain means the industry needs to focus on decarbonisation and de-methanisation. This can be done by identifying and reducing leakages. With increasingly unconventional sources of gas and more stringent decarbonisation targets, addressing CH_4 emissions is one clear way for the LNG industry to significantly reduce its emissions footprint.

Chapter summary and T-E-T-I analysis for the LNG industry

The T-E-T-I framework introduced in Chapter 2 and applied to the individual country analyses can also be applied to the LNG industry as a whole. Figure 6.11 summarises the state of the global LNG industry to conclude this chapter. LNG is the fastest-growing sector within natural gas, and liquid markets and standardisation

Resources

- Natural gas present in all major sections of the economy, and LNG is the fastest growing sector within natural gas
- LNG industry has access to abundant supplies including shale production, via proven reserves and new technology
- Recent natural gas and LNG infrastructure investments support market development

Market

- Liquid markets, standardi sation of contracts, and increased liquidity facilitate LNG trade
- LNG can drive coal-to-gas switching, and to an extent, oil switching, subject to overall affordability
- Renewable energy with battery storage is increasingly competitive
- Asia, a key market, is particularly sensitive to affordability and competitiveness

Energy-transition Dynamics

- Net-zero commitments put pressure on natural gas long term
- CCS is a key enabler to significantly decarbonis e the LNG chain, and natural gas' role in power generation
- CH_4 emissions, fugitive and flared, will have to be measured and reduced significantly with transparency
- Hydrogen can be an opportunity for LNG in the short term; longer term, renewable produced hydrogen will be a competitor

Power

- US tolling model, traders, and other participants decouple traditionally integrated value chain into various components
- Number of overall participants in the industry expanded over the past several decades
- IDCs and NOCs have heavily invested into LNG to lower carbon intensity and transition portfolios
- Capital intensive financial markets and the risk of stranded assets for newly financed projects

Figure 6.11 T-E-T-I analysis – LNG industry

of contracts have facilitated LNG trade. Net-zero commitments place natural gas under pressure in the long term, although hydrogen may be an opportunity for LNG in the short term. The number of participants in the LNG industry has expanded over the past few decades and seems set to expand further with new African players entering the market and filling the LNG gap in Europe created by the Ukraine/Russian war. Commercial flexibility will be needed as the US tolling model, traders, and other participants deconstruct the traditionally integrated value chain into smaller sub-components.

Note

1 GECF member countries include Algeria, Bolivia, Egypt, Equatorial Guinea, Iran, Libya, Nigeria, Qatar, Russia, Trinidad and Tobago, United Arab Emirates, and Venezuela, with observer countries of Angola, Azerbaijan, Iraq, Malaysia, Mozambique, Norway, and Peru.

References

Abe, A., Nakamura, M., Sato, I., Uetani, H., & Fujitani, T. (1998). Studies of the large-scale sea transportation of liquid hydrogen. *International Journal of Hydrogen Energy*, *23*(2), 115–121. https://doi.org/10.1016/S0360-3199(97)00032-3

Ajayi, T. A. (2024). Natural gas rents and institutions as co-growth drivers: Evidence from gas exporting countries forum with a panel 2SLS approach. *International Journal of Energy Sector Management*, *18*(3), 474–499. https://doi.org/10.1108/IJESM-01-2023-0015

Al Ghafri, S. Z. S., Revell, C., Di Lorenzo, M., Xiao, G., Buckley, C. E., May, E. F., & Johns, M. (2023). Techno-economic and environmental assessment of LNG export for hydrogen production. *International Journal of Hydrogen Energy*, *48*(23), 8343–8369. https://doi.org/10.1016/j.ijhydene.2022.11.160

Al-Kuwari, O. (2020, May 18). *Our new energy transformation can see the Middle East thrive – if it invests in RDI and education*. Qatar Foundation Op-Ed. www.qf.org.qa/stories/op-ed-our-new-energy-transformation-can-see-the-middle-east-thrive-if-it

Al-Kuwari, O. (2022). LNG's role in a decarbonising energy system: Lessons from the United Kingdom. In D. S. Olawuyi & E. G. Pereira (Eds.), *The Palgrave handbook of natural gas and global energy transitions* (pp. 141–171). Palgrave Macmillan.

Al-Kuwari, O. (2023). *The future of liquified natural gas (LNG) in the energy transition: Options and implications for the LNG industry in a decarbonising world* [Unpublished doctoral dissertation]. University College London. https://discovery.ucl.ac.uk/id/eprint/10173650/

Al-Kuwari, O., & Schönfisch, M. (2022). The emerging hydrogen economy and its impact on LNG. *International Journal of Hydrogen Energy*, *47*(4), 2080–2092. https://doi.org/10.1016/j.ijhydene.2021.10.206

Australian Renewable Energy Agency. (2018). *Hydrogen for Australia's future*. https://arena.gov.au/assets/2018/09/hydrogen-white-paper-finkel.pdf

Baker, R. (2020). *LNG finance: Will lenders accommodate the changing environment?* Oxford Institute for Energy Studies. www.oxfordenergy.org/wpcms/wp-content/uploads/2020/11/Insight-78-LNG-Finance-will-lenders-accommodate-the-changing-environment.pdf

Bang, Y., Seo, J. G., & Song, I. K. (2011). Hydrogen production by steam reforming of liquefied natural gas (LNG) over mesoporous Ni-La-Al_2O_3 aerogel catalysts: Effect of La content. *International Journal of Hydrogen Energy*, *36*(14), 8307–8315. https://doi.org/10.1016/j.ijhydene.2011.04.126

BEIS (Department of Business Energy & Industrial Strategy). (2020). *Greenhouse gas reporting: Conversion factors 2020*. www.gov.uk/government/publications/greenhouse-gas-reporting-conversion-factors-2020

Blanton, E., & Mosis, S. (2021). *The carbon-neutral LNG market: Creating a framework for real emissions reductions*. Columbia SIPA Center on Global Energy Policy. www.energypolicy.columbia.edu/wp-content/uploads/2021/07/Carbon-neutral-LNG-commentary-design-reflow-7.02.21-.pdf

Bose, A., Cohen, J., Fattouh, B., Johnson, O., & Spilker, G. (2021). *Voluntary markets for carbon offsets: Evolution and lessons for the LNG market*. Oxford Institute for Energy Studies. www.oxfordenergy.org/wpcms/wp-content/uploads/2021/10/Voluntary-markets-for-carbon-offsets-Evolution-features-and-lessons-for-the-LNG-market-ET03.pdf

BP (British Petroleum). (2020). *BP energy outlook 2020*. www.bp.com/content/dam/bp/business-sites/en/global/corporate/pdfs/energy-economics/energy-outlook/bp-energy-outlook-2020.pdf

BP (British Petroleum). (2021). *BP statistical review of world energy 2021*. www.bp.com/content/dam/bp/business-sites/en/global/corporate/pdfs/energy-economics/statistical-review/bp-stats-review-2021-full-report.pdf

Bresciani, G., Heiligtag, S., Lambert, P., & Rogers, M. (2020). *The future of liquefied natural gas: Opportunities for growth*. McKinsey & Company. www.mckinsey.com/industries/oil-and-gas/our-insights/the-future-of-liquefied-natural-gas-opportunities-for-growth

Bruce, S., Temminghoff, M., Hayward, J., Schmidt, E., Munnings, C., Palfreyman, D., & Hartley, P. (2018). *National hydrogen roadmap*. Commonwealth Scientific and Industrial Research Organisation. www.csiro.au/-/media/Do-Business/Files/Futures/18-00314_EN_NationalHydrogenRoadmap_WEB_180823.pdf

Caldecott, B., Howarth, N., & McSharry, P. (2013). *Stranded assets in agriculture: Protecting value from environment-related risks*. University of Oxford. www.smithschool.ox.ac.uk/sites/default/files/2022-03/stranded-assets-agriculture-report-final.pdf

Egging, R., & Tomasgard, A. (2018). Norway's role in the European energy transition. *Energy Strategy Reviews*, *20*, 99–101. https://doi.org/10.1016/j.esr.2018.02.004

Energy Capital & Power. (2024, May 14). *IAE 2024 shines spotlight on Africa's gas future*. https://energycapitalpower.com/iae-2024-shines-spotlight-on-africas-gas-future/

European Commission. (2019). *The European green deal COM 2019 (640 final 2)*. www.eea.europa.eu/policy-documents/com-2019-640-final

European Commission. (2021, October 31). *International methane emissions observatory launched to boost action on powerful climate-warming gas*. https://ec.europa.eu/commission/presscorner/detail/es/ip_21_5636

European Commission. (2022). *European taxonomy: Complementary climate delegated act to accelerate decarbonisation*. https://ec.europa.eu/commission/presscorner/detail/en/ip_22_711#:~:text=The%20aim%20of%20the%20EU,Europe%20climate%20neutral%20by%202050

European Commission. (2024, January 16). *In focus: Methane emissions*. https://energy.ec.europa.eu/news/focus-methane-emissions-2024-01-16_en#:~:text=EU%20actions%20on%20methane&text=The%20Commission%20published%20its%20Methane,as%20the%20production%20of%20biomethane

Fong, W. K., Sotos, M., Doust, M., Shultz, S., Marques, A., & Deng-Beck, C. (2021). *Global protocol for gas emission inventories – an accounting and reporting standard for cities, version 1.1*. Greenhouse Gas Protocol. https://ghgprotocol.org/sites/default/files/standards/GPC_Full_MASTER_RW_v7.pdf

Fulwood, M., Henderson, J., Gomes, I., Lambert, M., & Sharples, J. (2020). *Emerging Asia LNG demand*. Oxford Institute for Energy Studies. www.oxfordenergy.org/wpcms/wp-content/uploads/2020/09/Emerging-Asia-LNG-demand-NG-162.pdf

Gabbatiss, J. (2020, April 6). *The carbon brief profile: South Korea*. CarbonBrief. www.carbonbrief.org/the-carbon-brief-profile-south-korea/

GECF (Gas Exporting Countries Forum). (2023). *GECF annual statistical bulletin*. www.gecf.org/_resources/files/events/release-of-the-gecf-annual-statistical-bulletin-2023/gecf-asb-2023.pdf

GECF (Gas Exporting Countries Forum). (2024). *Annual gas market report 2024*. www.gecf.org/_resources/files/events/gecf-unveils-the-annual-gas-market-report-2024/gecf-agmr-2024.pdf

GIIGNL (The International Group of Liquefied Gas Importers). (2020). *LNG insight – LNG carbon offsetting: Fleeting trend or sustainable practice?* https://nasslng.ru/assets/files/giignl_lng_carbon_offsetting_061820.pdf

GIIGNL (The International Group of Liquefied Gas Importers). (2021). *MRV and GHG neutral LNG framework*. https://giignl.org/wp-content/uploads/2021/11/MRV-and-GHG-Neutral-Framework.pdf

Gilbert, A. Q., & Sovacool, B. K. (2017). US liquefied natural gas (LNG) exports: Boom or bust for the global climate? *Energy*, *141*, 1671–1680. https://doi.org/10.1016/j.energy.2017.11.098

Global Energy Association. (2023, August 10). *Qatar could increase its LNG exports by 60%*. https://globalenergyprize.org/en/2023/08/10/qatar-could-increase-its-lng-exports-by-60/

Greenpeace. (2023). *The mirage of emission reduction: The risks of using forestry carbon offset credits in the oil and gas industry*. www.greenpeace.org/static/planet4-eastasia-stateless/2023/12/22b1909b-the-mirage-of-emission-reduction.pdf

Grubb, M., Hourcade, J. C., & Neuhoff, K. (2015). The three domains structure of energy-climate transitions. *Technological Forecasting and Social Change*, *98*, 290–302. https://doi.org/10.1016/j.techfore.2015.05.009

Gustafson, T. (2020). *The bridge: Natural gas in a redivided Europe*. Harvard University Press.

Hanley, E. S., Glowacki, B. A., Nuttall, W. J., & Kazantzis, N. (2015). Natural gas – synergies with hydrogen. *Proceedings of the Institution of Civil Engineers – Energy*, *168*(1), 47–60. https://doi.org/10.1680/ener.14.00018

ICAP (International Carbon Action Partnership). (2022). *China national ETS*. https://icapcarbonaction.com/system/files/ets_pdfs/icap-etsmap-factsheet-55.pdf

IEA (International Energy Agency). (2019). *The future of fuel: The future of hydrogen*. IEA Publications. www.iea.org/reports/the-future-of-hydrogen

IEA (International Energy Agency). (2020). *World energy investment 2020*. IEA Publications. https://doi.org/10.1787/6f552938-en

IEA (International Energy Agency). (2021a). *Gas market report Q3–2021*. IEA Publications. www.iea.org/reports/gas-market-report-q3-2021

IEA (International Energy Agency). (2021b). *Net zero by 2050. A roadmap for the global energy sector*. IEA Publications. www.iea.org/reports/net-zero-by-2050

IEA (International Energy Agency). (2023a). *Global hydrogen review 2023*. IEA Publications. https://iea.blob.core.windows.net/assets/ecdfc3bb-d212-4a4c-9ff7-6ce5b1e19cef/GlobalHydrogenReview2023.pdf

IEA (International Energy Agency). (2023b). *World energy investment 2023*. IEA Publications. https://iea.blob.core.windows.net/assets/8834d3af-af60-4df0-9643-72e2684f7221/WorldEnergyInvestment2023.pdf

IEA (International Energy Agency). (2023c). *World energy outlook 2023*. IEA Publications. https://iea.blob.core.windows.net/assets/86ede39e-4436-42d7-ba2a-edf61467e070/WorldEnergyOutlook2023.pdf

IEA (International Energy Agency). (2024a). *Batteries and secure energy transitions*. IEA Publications. https://iea.blob.core.windows.net/assets/cb39c1bf-d2b3-446d-8c35-aae6b-1f3a4a0/BatteriesandSecureEnergyTransitions.pdf

IEA (International Energy Agency). (2024b). *Gas market report, Q2–2024*. IEA Publications. https://iea.blob.core.windows.net/assets/6bc34660-dd8e-4018-bb21-973023846b4c/GasMarketReport%2CQ2-2024.pdf

IEA (International Energy Agency). (n.d.). *Asia Pacific*. Retrieved July 15, 2024, from www.iea.org/regions/asia-pacific/energy-mix

IGU (International Gas Union). (2021a). *World LNG report 2021*. www.igu.org/resources/world-lng-report-2021/

IGU (International Gas Union). (2021b, October 28). *Getting it right to reduce methane emissions*. https://igu.org/news/getting-it-right-to-reduce-methane-emissions/

Liu, H., Evans, S., Zhang, Z., Song, W., & You, X. (2023, November 30). *The carbon brief profile: China*. CarbonBrief. https://interactive.carbonbrief.org/the-carbon-brief-profile-china/

Lloyd's Register & UMAS. (2019). *Zero-emission vessels: Transition pathways*. www.lr.org/en/knowledge/press-room/press-listing/press-release/lr-and-umas-release-new-zero-emission-vessels-transition-pathways-study/

McHich, A., & Britnell, C. (2021, June 10). *Going green with carbon-neutral LNG*. CME Group. www.cmegroup.com/education/articles-and-reports/going-green-with-carbon-neutral-lng.html

Mitrova, T., Melnikov, Y., & Chugunov, D. (2019). *The hydrogen economy – a path towards low carbon development*. SKOLKOVO Moscow School of Management. https://doi.org/10.13140/RG.2.2.23559.75686

Ouki, M. A. (2023). *Prospects for a potential African gas renaissance en route to a "just energy transition"*. Oxford Institute of Energy Studies. https://hdl.handle.net/10419/280128

Procaccini, G., & Markowitz, K. (2020, February 1). *Key considerations for carbon-neutral oil and LNG transactions using carbon offsetts*. Akin Gump Strauss Hauer & Feld LLP. www.akingump.com/en/experience/industries/energy/speaking-energy/key-considerations-for-carbon-neutral-oil-and-lng-transactions-using-carbon-offsets.html

Qadrdan, M., Chaudry, M., Jenkins, N., Baruah, P., & Eyre, N. (2015). Impact of transition to a low carbon power system on the GB gas network. *Applied Energy*, *151*(2015), 1–12. https://doi.org/10.1016/j.apenergy.2015.04.056

Roman-White, S., Rai, S., Littlefield, J., Cooney, G., & Skone, T. J. (2019). *Life cycle greenhouse gas perspective on exporting liquefied natural gas from the United States: 2019 update*. National Energy Technology Laboratory. www.osti.gov/biblio/1607677

Royal Dutch Shell. (2021). *Strategic report*. https://reports.shell.com/annual-report/2021/strategic-report.html

Staffell, I., Scamman, D., Abad, A. V., Balcombe, P., Dodds, P. E., Ekins, P., Shah, N., & Ward, K. R. (2018). The role of hydrogen and fuel cells across the global energy system. *Energy & Environmental Science*, *2*, 463–491. https://doi.org/10.17605/OSF.IO/RZM4G

Stern, J. (2017). *Challenges to the future of gas: Unburnable or unaffordable?* Oxford Institute for Energy Studies. www.oxfordenergy.org/wpcms/wp-content/uploads/2017/12/Challenges-to-the-Future-of-Gas-unburnable-or-unaffordable-NG-125.pdf

Stern, J. (2019a). *Challenges to the future of LNG*. Oxford Institute for Energy Studies. www.oxfordenergy.org/wpcms/wp-content/uploads/2019/10/Challenges-to-the-Future-of-LNG-NG-152.pdf

Stern, J. (2019b). *Narratives for natural gas in decarbonising European energy markets*. Oxford Institute for Energy Studies. www.oxfordenergy.org/wpcms/wp-content/uploads/2019/02/Narratives-for-Natural-Gas-in-a-Decarbonisinf-European-Energy-Market-NG141.pdf

Stern, J. (2020). *Methane emissions from natural gas and LNG imports: An increasingly urgent issue for the future of gas in Europe*. Oxford Institute for Energy Studies. www.oxfordenergy.org/wpcms/wp-content/uploads/2020/11/Methane-Emissions-from-Natural-Gas-and-LNG-Imports-an-increasingly-urgent-issue-for-the-future-of-gas-in-Europe-NG-165.pdf

Steuer, C. (2019). *Outlook for competitive LNG supply*. Oxford Institute for Energy Studies. https://doi.org/10.26889/9781784671310

Teplin, C., Dyson, M., Engel, A., & Glazer, G. (2019). *The growing market for clean energy portfolios: Economic opportunities for a shift from new gas-fired generation to clean energy*. RMI. https://rmi.org/cep-reports

Timera Energy. (2021, March 29). *The value impact of LNG carbon emissions*. https://timera-energy.com/blog/the-value-impact-of-lng-carbon-emissions/#:~:text=Cargo%20differentiation%20by%20carbon%20footprint,low%20carbon%20LNG'%20deals%20signed.

Total Energies. (2022). *Total energies 2021 – results & 2022 outlook*. https://totalenergies.com/sites/g/files/nytnzq121/files/documents/2022-02/220210_Master_Presentation_2021_Results_2022_Oultook.pdf

Van de Graaf, T., Overland, I., Scholten, D., & Westphal, K. (2020). The new oil? The geopolitics and international governance of hydrogen. *Energy Research and Social Science*, *70*, 101667. https://doi.org/10.1016/j.erss.2020.101667

Wood, D. A. (2024). Carbon-neutral LNG cargoes: A potentially valuable concept requiring improved transparency. In D. A. Wood & J. Cai (Eds.), *The fundamentals and sustainable advances in natural gas science and engineering* (pp. 445–475). Elsevier. https://doi.org/10.1016/B978-0-443-13420-3.00015-9

Wood Mackenzie. (2020a, August 17). *Renewable energy could reduce Asia Pacific LNG plant emissions by 8%* [Press release]. www.woodmac.com/press-releases/renewable-energy-could-reduce-asia-pacific-lng-plant-emissions-by-8/

Wood Mackenzie. (2020b, November 18). *What is carbon-neutral LNG*. www.woodmac.com/news/opinion/what-is-carbon-neutral-lng/

World Bank. (2020). *Global initiative to reduce global gas flaring: "Zero routine flaring by 2030"*. www.worldbank.org/zeroroutineflaring

World Energy Council. (2021). *World energy trilemma index*. www.worldenergy.org/assets/downloads/WE_Trilemma_Index_2021.pdf?v=1634811254

7 Conclusion

Global LNG volumes will grow substantially in the coming decade, in some scenarios more than doubling from 400 BCM in 2017 to almost 900 BCM in 2040 (BP, 2019). The global LNG market is supported by the overall growth in production of natural gas, which has reduced prices and increased consumption (Boersma & Jordaan, 2017; GECF, 2023; IEA, 2023b; Ouki, 2023). From a demand perspective, Asia and Europe combined could continue to dominate the LNG market in the coming decades, supported by strong demand arising from air-quality concerns and continuing gas-to-coal switching, and increased import dependency in Europe (BP, 2019; Equinor, 2019; IEA, 2016, 2023a). Supported by abundant gas supplies, the United States is forecast to be the fastest-growing supplier and seems set to become one of the four major exporting regions by 2040, alongside the Middle East, Asia Pacific, and Africa. This concluding chapter presents practical recommendations for LNG stakeholders: How can we ensure cleaner technologies and production processes? How can we develop the capacity to fully maximise the potential of LNG as a bridge fuel? The book introduced a new meta-framework – the T-E-T-I framework (Chapter 2) – tailored to energy transitions. Thus, this chapter also considers how the framework can guide such decision-making and transformation in the LNG market.

Global review and predictions

Increased LNG supplies will lead to greater competition globally between LNG and pipeline gas (BP, 2019; IEEFA, 2024). LNG is forecast to overtake interregional pipelines in volume by 2030 with overall capacity expected to double, while there are few prospects for new interregional pipeline projects (Corbeau & Ledesma, 2016; IEA, 2017, 2023b; BP, 2019). Increased LNG from the United States will add to the overall supply while also providing LNG from a different geographical region, thus applying pressure on traditional suppliers to Europe and Asia, a process that could support the "depoliticization" of the industry. With limited LNG expansions planned for Australia and Russia (for different reasons), there will ultimately be two major LNG exporters. On one hand, the United States could focus on delivering more flexibility and optionality, while Qatar focuses on delivering security and reliability. Some authors have also described an African LNG

DOI: 10.4324/9781003545149-7

"Renaissance" with many African countries developing their LNG supply potential and African policymakers using the European gas crisis to highlight the lack of international hydrocarbon investment into the African region (Ouki, 2023).

The UK and Japan have developed energy systems and are at similar stages in their energy transition. They are moving towards transforming their infrastructure via purposive measures. China's energy system is growing rapidly. Coal and nuclear energy use are variables that impact the evolution of these systems. Renewable energy is growing rapidly and is forecast to continue to do so, subject to geographical and market constraints. The Japanese and Chinese gas and power markets are less market-oriented than is the case in the UK. In Japan and China, government and vested interests play a relatively large role. In the UK, the Government supports a market-based approach to respond to supply-and-demand fundamentals. In terms of natural gas, the UK and China have access to both LNG and pipeline gas. Japan imports all its natural gas via LNG. The three countries analysed in detail in Chapters 3–5 have adopted very different approaches to climate change politics – as all countries do – reflecting their political and social dynamics.

How can we ensure cleaner technologies and production processes?

LNG appears set to maintain a major role in current and new gas markets until 2035. Asia – mainly China and India – will be the key long-term markets for the industry. Coal-to-gas and affordability of LNG in those markets will largely determine the viability of LNG beyond 2030. In Europe, LNG will support the energy transition by meeting demand from declining domestic production and ensuring security of supply. However, demand for LNG will gradually decline as decarbonisation policies take shape. LNG differs from pipeline gas. Through industrial processing, the LNG industry transforms CH_4 into a more flexible and potentially cleaner product that could provide an advantage when it comes to meeting the evolving market dynamics of the energy transition (Ajayi, 2024; GECF, 2023). Within the LNG industry, exporters will compete when it comes to cost, emissions, and reliability. LNG exporters' ability to reduce emissions on delivered LNG, increased transparency, and commercial flexibility can give suppliers a competitive advantage. The LNG industry will need to continue to decarbonise its value chain via commercial, technical, and operational methods.

Commercial methods may include trading or project structuring. In terms of trading, carbon-neutral cargoes are viewed as a bridging mechanism with the carbon-offset LNG market expected to exceed 25 million tonnes by 2030, subject to compliance with relevance offset principles. A cross-value chain downstream market has been developed for carbon-offset LNG in Australia's North Asian customer countries, including Japan (Ramesh & Bogle, 2023). In terms of project restructuring, LNG value chains can leverage certain geographies and industrial access via LNG terminals and incorporate projects with potential hydrogen uses (e.g. CH_4 reforming and pyrolysis). Technical aspects may include replacing LNG feedstock with low synthetic gas and using LNG infrastructure for "green LNG". This, however, depends on the availability of alternative gas (e.g. biomethane).

Upstream exports could be fully decarbonised by converting natural gas to blue hydrogen through reforming or pyrolysis; this approach depends on CCS availability and requires significant investment. Operationally, full value chain emission measurement and certification are key areas of focus by the industry. However, unified industry standards have not yet been developed and there remains a lack of transparency regarding emissions data.

Globally, the uncertainties of the energy transition and its dynamics have the potential to expand among the historically concentrated group of exporters and importers in the LNG industry. The recent crisis in Europe has pushed Germany and other countries to become active players in Europe to diversify from Russian gas. Asian LNG demand is likely to increase to mitigate coal use and support renewables, including new importers such as Vietnam and the Philippines. While Latin America is a small player currently, Brazil, Argentina, and other countries can play an increasingly important seasonal role as demand continues to gradually increase. Increased global demand has spurred investment in LNG projects, not just from Qatar and the United States, but also from multiple African players (e.g. Mozambique, Ghana, Tunisia, Cote d'Ivoire, and Cameroon) that will also look to access traditional and emerging markets (Energy Capital & Power, 2024; IEEFA, 2024; IGU, 2024; Ouki, 2023). The last seven years have seen the development of several large natural gas projects in sub-Saharan Africa. These LNG export projects include Coral Sul FLNG (commissioned in November 2022) and Mozambique LNG in Mozambique; the Grand Tortue Ahmeyim (GTA) LNG project in Mauritania – Senegal, and Ghana's Sankofa gas development project or Offshore Cape Three Points (OCTP) project (IGU, 2024; Ouki, 2023). With advancements in technology, and different energy-transition dynamics, LNG's footprint could increase in the coming decades.

How can we develop the capacity to fully maximise the potential of LNG as a bridge fuel?

Intermittency, energy security, and overall system costs could play in favour of LNG. Intermittency is a significant factor supporting the ongoing need for large dispatchable electricity sources, at least until renewables and batteries reach global scale. Key enablers for the future of LNG include CCS, short- to medium-term hydrogen demand, and carbon and CH_4 management. Without successful scale-up of CCS, there will be a limited future for natural gas beyond the transition fuel bridging period (anticipated to occur circa 2030/2035). Natural gas-based hydrogen demand could support LNG demand in the short to medium term. However, the LNG industry does not have the ability to transform itself into a hydrogen industry in the long-term. Renewables, green hydrogen, and decarbonisation targets will continue to put pressure on LNG. The role of gas in more mature economies will decrease with the increase of renewable power and the introduction of alternative methods of heating. Green hydrogen will likely displace blue hydrogen and natural gas in key sectors. If LNG cannot compete in Asia (particularly with coal), and if green hydrogen and renewable energy technology accelerate, then key markets may "skip the gas step", naturally hindering the LNG industry.

Application of the T-E-T-I framework

The T-E-T-I framework introduced in Chapter 2 and applied to the country cases in subsequent chapters is based on institutional, social, and techno-economic theories. Through its application, three interconnected themes were identified to determine the role of natural gas usage and LNG in the energy transition: (1) energy security, (2) climate change and environmental policies, and (3) affordability and competitiveness. Table 7.1 summarises these three main areas, their linked sub-themes and key concepts. This book provided insights into the relationship between the above-mentioned three core themes. These relationships translate into policy priorities and affect decisions related to LNG use in different energy systems. In the context of energy-transition dynamics, three LNG pathway types were identified:

1. The transformation pathway, where energy security and climate change are strongly linked.
2. The transition fuel pathway, where energy security is the prominent policy goal.
3. Integrated development pathway, where all three priorities are important, with affordability and energy security taking slight priority.

Countries assessing options for their fuel mix must consider many variables, and how these are prioritised and factored into decision-making will determine the

Table 7.1 Key themes and sub-themes of case study analysis

Theme	*Sub-theme*	*Key concepts*
Energy security	Decarbonisation impact on energy security	Domestic low-carbon resources Import dependence
	LNG challenges to energy security	Supply availability (during demand peaks) Trade imbalance Geopolitical risk (disruptions)
	LNG measures to support energy security (mitigation)	Resource diplomacy Upstream investment Source diversification
Climate change and environmental policies	Global climate politics	Climate leadership/Paris Agreement alignment Net-zero targets
	Natural gas narratives	Transition fuel Emissions
	Decarbonisation pathways	Abatement curves Hard-to-abate sectors Molecules versus electrons
Affordability and competitiveness	Affordability as a policy driver	Market liberalisation Liquid market Volatility exposure
	Economics of decarbonisation	Transition costs Carbon price Competing sources

LNG pathway followed. Taking into consideration the conditions for natural gas to be a successful transition fuel, the key criteria will be timing, climate ambition, CH_4 emissions, substitution, carbon lock-in, and the availability of CCS technology. Table 7.2 summarises the critical factors in the energy-transition context.

The LNG industry analysis presented in this text categorises challenges and opportunities into three key areas: (1) technology and alternatives, (2) rapid transition, and (3) evolving market dynamics. Taking into consideration the three pathways identified above, there is a clear role for LNG during the energy transition. The domain 1 "satisfice" approach (with elements of domains 2 and 3, Grubb et al., 2015) can be used for the integrated development pathway, with the following practical recommendations for stakeholders:

- *Meet growing demand*: LNG has the potential to support growth in fast-growing economies. This can be done through "plug and play" and scalable approaches,

Table 7.2 Success factors for natural gas as a transition fuel based on Al-Kuwari (2023)

Criteria	*Domain*	*Consideration*	*Impact*
Timing	Economics Environmental Political	Certainty of demand for natural gas	Competing options for natural gas, imports versus local production
Climate ambition	Environmental Political Economics	Driven by policy driver – decarbonisation – fuel-switching	Driver for climate ambition may determine long- versus short-term strategy
CH_4 emissions	Environmental Political	Position of gas globally (international agreements) Technologies and agreements to mitigate emissions	CH_4 emissions management will affect the long-term viability of natural gas
Fuel-switching	Economics Environmental	Imported gas – replace higher carbon sources – reduce risk of incremental overall demand	Increase in overall emissions if natural gas does not replace higher carbon options
Carbon lock-in	Economics Political	Infrastructure development – utilise current assets – minimise investment – reduce risk of stranded assets	Risk of unutilised public investment
CCS	Economics Political Environmental	CCS options – availability of sites – public acceptance of CCS	Availability and access will determine long-term strategy

such as LNG-to-power. Increased natural gas supply contributes to increased LNG supply and to suppliers providing additional options for LNG importers. The increasing number of players involved supports a growing spot market, reducing the need to enter long-term contracts.

- *Support coal-to-gas and oil-to-gas switching*: LNG can replace coal rapidly and at significant scale, while reducing emissions. This is particularly the case for countries where power demand exceeds the ability to develop low-carbon electricity, thereby often positioning LNG as the most suitable solution. This must be supported by increasing transparency and verification of LNG emissions across the chain. The LNG industry can also target key industrial sectors that require high heat (e.g. steel, food production, and ceramics).
- *Competitive economics*: The LNG industry can explore ways to lower overall supply costs. Low-cost LNG projects – especially those with low-carbon footprints – will be successful in the medium to long term. Once an LNG project is funded, there are relatively low marginal costs of supply.
- *Improve "transition fuel" communication and narratives*: The LNG industry has an opportunity to establish stronger narratives around the important role it plays (compared to pipeline gas and similar commodities). This is particularly the case when it comes to energy security.
- *Address environmental footprint*: Increased consumer awareness and consumer choice will allow for comparing LNG to alternatives (including imported pipeline gas). Sources of LNG and upstream operations (involving flaring and CH_4 emissions) will be increasingly scrutinised, at least until shale gas GHG emission profiles are better understood.

In integrated development pathways, policymakers can take advantage of increased LNG supply to meet growing demand as countries move away from coal. Coal-to-gas switching can be encouraged via carbon pricing mechanisms and financial support for new importing infrastructures.

The domain 2 "optimise" approach can be applied to the transition fuel pathway, with the following recommendations:

- *Intermittency, backup, and seasonality*: In developed markets, renewable electricity is increasing rapidly. The LNG industry could develop solutions that support intermittency as a source of baseload power until long duration storage competes economically. LNG can also be used to meet seasonal demands, mainly related to heating.
- *Technical flexibility*: The industry can focus on the development of technological solutions such as FSRUs and floating LNG, allowing for "plug and play" options. Plug and play options bring on LNG rapidly to meet demand during stagnant domestic production or as a bridge to alternative solutions. Flexibility around LNG import technology and LNG-to-power solutions minimises the lock-in risks of alternatives, including complex pipeline gas projects.
- *Resilience*: Different types of LNG storage need to be developed to support additional resilience in energy systems. This is witnessed during extreme weather conditions (e.g. the 2018 winter in Europe and the 2021 Texas power crisis).

- *Established infrastructure*: Promoting the ability to leverage and use established gas infrastructure could support lower-cost decarbonisation pathways by using CCS.
- *Use commercial tools to support net-zero*: Encourage carbon-neutral cargoes and the use of carbon offsets to meet importer needs.
- *Decarbonising the value chain*: The LNG industry can take proactive measures to increase transparency and reduce emissions through improved communication.

In transition pathways, policymakers can promote LNG as a support for the transition to renewables and as backup for seasonal purposes. Market access mechanisms must be designed to ensure that the spot and short-term markets are dynamic enough to meet demand and energy security requirements.

The domain 2 and 3 "optimise and transform" approaches can be used for the transformation pathway. Relevant recommendations include the following:

- *Decarbonising and "demethanising" the value chain*: The LNG industry can take aggressive measures to begin decarbonising the value chain and reduce overall emissions through technological and operational improvements. Carbon-neutral cargoes, emission offsets, and other commercial measures are attempts to ensure that LNG is well positioned in the short to medium term.
- *Explore exporting hydrogen-based products*: This can initially be done via LNG in industrial clusters and hydrogen carriers like ammonia, an approach that transforms the product from LNG to CH_4. This is feasible in cases that dependent on geography, value chain exposure, and market dynamics.
- *Transform via innovation*: Use a combination of current and future technologies (e.g. CCS, hydrogen, and pyrolysis) beyond the narrow value chain approach. LNG players can invest in innovation for a comprehensive net-zero approach.

In transformation pathways, policymakers can leverage natural gas infrastructure as a bridge to low-carbon systems. This can be done by focusing on seasonality, intermittency (short-term), and transitioning to renewables and hydrogen-based economies (long-term).

A purely development-first pathway, where the focus is efficiency and rapidly meeting demand, could also be implemented in developing economies, such as the Philippines, India, and African countries where affordability is the priority. The need for gas infrastructure and the ability of LNG to be competitive must be seriously considered.

The emerging hydrogen economy and its impact on the development of LNG

Hydrogen can be deployed in long-haul transportation, industry, and heating (both directly and via blending). Gas-based low-carbon hydrogen gas demand could support LNG demand in pathways where renewable energy-based hydrogen is not competitive. High levels of gas-based low-carbon hydrogen production have the potential to stabilise global LNG demand in the medium to long term.

However, LNG might be threatened by the development of a hydrogen economy if (a) renewable-based low-carbon energy dominates for cost or policy-related reasons or (b) CCS cannot be deployed at scale. Commercial and operational synergies could assist the LNG industry to participate in and develop a value chain around natural-gas-based low-carbon hydrogen. Figure 7.1 illustrates synergies in (1) a natural-gas, low-carbon hydrogen value chain (blue hydrogen) and (2) a renewable energy-based low-carbon hydrogen value chain (green hydrogen). In both, liquefaction capacity, receiving terminals, and pipeline grids can be utilised to varying degrees. Hydrogen carriers, such as ammonia, could also provide various pathways for upstream gas, but elements of the LNG value chain could also play a role.

The fundamental difference between pipeline gas and LNG is market access. Pipeline gas is intrinsically regional and economically limited to several thousand kilometres. It is also subject to different configurations; both exporters and importers are locked into a limited set of options. LNG provides access to a global market, which benefits all participants in the value chain. Market dynamics are very different, notably when it comes to pricing and commercial terms. While CH_4 is the end-product for both pipeline gas and LNG, it requires significant capital investment to liquefy and ship LNG. LNG has more capacity to address carbon lock-in and to adjust to global market needs.

LNG is currently favoured over pipeline gas. There is geopolitical sensitivity around major transnational pipelines, and LNG players have displayed technical flexibility and the ability to rapidly build LNG infrastructure. LNG is often seen as a replacement for pipeline gas, mainly because there are no efficient lower-carbon alternatives. This has led to an increase in coal use and increased interest in nuclear energy. The economic impact and the energy security imperative could also garner public support for a move away from natural gas, perhaps, at an even faster trajectory. That said, the energy crisis has accelerated investment in LNG infrastructure, partly backed by long-term contracts. In Europe, there is now additional LNG infrastructure, supported by gas infrastructure (storage and interconnections). This means that the issue is not where there will be further gas infrastructure investment, but how it will be used going forward. This could present opportunities for LNG exporters and importers moving forwards. The T-E-T-I framework introduced in this text is valuable for integrating complex variables from multiple domains for the construction of a holistic picture – essential to make reasoned decisions about LNG's role in the energy transformation in specific (e.g. country), regional, or broader global settings.

It remains to be seen what impact the energy crisis will have on the future of LNG. The challenges introduced by this sudden disruption highlight the need for flexibility and resilience. It does though highlight the benefits of natural gas as a stable source of baseload energy. The public have been able to observe vulnerabilities in energy systems in real time, and the crisis has spurred some additional measures to move away from gas as a transition fuel. LNG's future is broadly linked to the role of natural gas in future energy systems. However, LNG's development has its own trajectory when compared to pipeline gas and other fossil fuels as explained in this text.

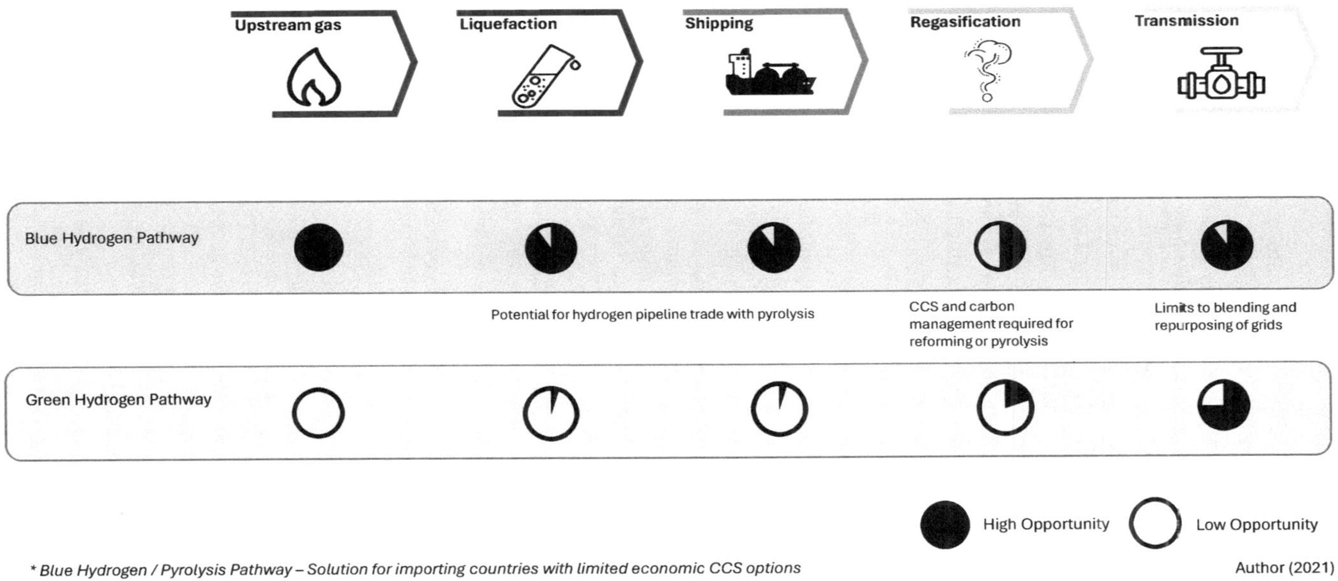

Figure 7.1 Technical low-carbon hydrogen and LNG value chain synergies

Energy transitions are periods of change and opportunity, and are not new. Previous generations experienced the Industrial Revolution, or the "age of steam". To make our economies and societies resilient in the face of the current low-carbon energy transition – driven largely by climate change and its impact on our planet – we have to invest in research, development, and innovation (RDI) and education. Al-Kuwari (2020), writing specifically about the role of the Middle East in the energy transformation, noted that "to thrive long-term, a serious focus on RDI and education – combined with economic diversification – will be key" (p. 1). The current energy transition is especially critical given the race against the world's carbon budget. The impact of COVID-19 – reflected in energy markets by the collapse in the price of US oil futures – indicates just how rapidly our world can change. It also demonstrates why developing a thriving RDI and education ecosystem is the best – or possibly only – way forward to build resilient and adaptable societies and economies that are ready to manage the upcoming global low-carbon transition.

References

Ajayi, T. A. (2024). Natural gas rents and institutions as co-growth drivers: Evidence from gas exporting countries forum with a panel 2SLS approach. *International Journal of Energy Sector Management, 18*(3), 474–499. https://doi.org/10.1108/IJESM-01-2023-0015

Al-Kuwari, O. (2020, May 18). *Our new energy transformation can see the Middle East thrive – if it invests in RDI and education.* Qatar Foundation Op-Ed. www.qf.org.qa/stories/op-ed-our-new-energy-transformation-can-see-the-middle-east-thrive-if-it

Al-Kuwari, O. (2023). *The future of liquified natural gas (LNG) in the energy transition: Options and implications for the LNG industry in a decarbonising world* [Unpublished doctoral dissertation]. University College London. https://discovery.ucl.ac.uk/id/eprint/10173650/

Boersma, T., & Jordaan, S. M. (2017). Whatever happened to the golden age of natural gas? *Energy Transitions, 1*(2), 1–8. https://doi.org/10.1007/s41825-017-0005-4

BP (British Petroleum). (2019). *BP energy outlook 2019.* www.bp.com/content/dam/bp/business-sites/en/global/corporate/pdfs/energy-economics/energy-outlook/bp-energy-outlook-2019.pdf

Corbeau, A., & Ledesma, D. (Eds.). (2016). *LNG markets in transition: The great reconfiguration.* Oxford University Press for the Oxford Institute for Energy Studies and the King Abdullah Petroleum Studies and Research Center.

Energy Capital & Power. (2024, May 14). *IAE 2024 shines spotlight on Africa's gas future.* https://energycapitalpower.com/iae-2024-shines-spotlight-on-africas-gas-future/

Equinor. (2019). *Energy perspectives 2019: Long-term macro and market outlook.* www.equinor.com/sustainability/energy-perspectives-previous-reports

GECF (Gas Exporting Countries Forum). (2023). *GECF annual statistical bulletin.* www.gecf.org/_resources/files/events/release-of-the-gecf-annual-statistical-bulletin-2023/gecf-asb-2023.pdf

Grubb, M., Hourcade, J. C., & Neuhoff, K. (2015). The three domains structure of energy-climate transitions. *Technological Forecasting and Social Change, 98*, 290–302. https://doi.org/10.1016/j.techfore.2015.05.009

IEA (International Energy Agency). (2016). *Energy and air pollution – world energy outlook: Special report 266.* IEA Publications. https://doi.org/10.1021/ac00256a010

IEA (International Energy Agency). (2017). *World energy investment 2018.* IEA Publications. https://doi.org/10.1787/energy_tech-2017-en

IEA (International Energy Agency). (2023a). *Outlooks for gas markets and investment.* IEA Publications. https://iea.blob.core.windows.net/assets/5cce10d6-6c88-4813-a40f-ceffecdb0986/Outlooksforgasmarketsandinvestment.pdf

IEA (International Energy Agency). (2023b). *World energy outlook 2023*. IEA Publications. https://iea.blob.core.windows.net/assets/86ede39e-4436-42d7-ba2a-edf61467e070/WorldEnergyOutlook2023.pdf

IEEFA (Institute for Energy Economics and Financial Analysis). (2024). *Global LNG outlook 2024–2028*. https://ieefa.org/sites/default/files/2024-04/Global%20LNG%20Outlook%202024-2028_April%202024%20%28Final%29.pdf

IGU (International Gas Union). (2024). 2024 *World LNG report*. www.igu.org/resources/2024-world-lng-report/?gad_source=1&gclid=CjwKCAjwtNi0BhA1EiwAWZaANGrxBEijHaSFlx_4aUrfEObTjr-P57PnTxnydkwkevk1XHFXIG1_pRoC1-oQAvD_BwE

Ouki, M. A. (2023). *Prospects for a potential African gas renaissance en route to a "just energy transition"*. Oxford Institute of Energy Studies. https://hdl.handle.net/10419/280128

Ramesh, K., & Bogle, S. (2023). Under the loupe: The case for carbon-offset LNG from Australia. *APPEA Journal*, *63*(2), S428–S433. https://doi.org/10.1071/AJ22211

Index

Note: page numbers and page spans for figures indicated in *italics*; tables in **bold**

FSC
www.fsc.org
MIX
Papier | Fördert
gute Waldnutzung
FSC® C083411

Kolibri 360 GmbH
Ferdinand-Jühlke-Straße 7
99095 Erfurt, Deutschland
produktsicherheit@kolibri360.de